Image Processing and Acquisition using Python

CHAPMAN & HALL/CRC MATHEMATICAL AND COMPUTATIONAL IMAGING SCIENCES

Series Editors

Chandrajit Bajaj
Center for Computational Visualization
The University of Texas at Austin

Guillermo Sapiro
Department of Electrical
and Computer Engineering
Duke University

Aims and Scope

This series aims to capture new developments and summarize what is known over the whole spectrum of mathematical and computational imaging sciences. It seeks to encourage the integration of mathematical, statistical and computational methods in image acquisition and processing by publishing a broad range of textbooks, reference works and handbooks. The titles included in the series are meant to appeal to students, researchers and professionals in the mathematical, statistical and computational sciences, application areas, as well as interdisciplinary researchers involved in the field. The inclusion of concrete examples and applications, and programming code and examples, is highly encouraged.

Published Titles

Image Processing for Cinema
by Marcelo Bertalmío

Image Processing and Acquisition using Python
by Ravishankar Chityala and Sridevi Pudipeddi

Statistical and Computational Methods in Brain Image Analysis
by Moo K. Chung

Rough Fuzzy Image Analysis: Foundations and Methodologies
by Sankar K. Pal and James F. Peters

Theoretical Foundations of Digital Imaging Using MATLAB®
by Leonid P. Yaroslavsky

Proposals for the series should be submitted to the series editors above or directly to:
CRC Press, Taylor & Francis Group
3 Park Square, Milton Park, Abingdon, OX14 4RN, UK

CHAPMAN & HALL/CRC
MATHEMATICAL AND COMPUTATIONAL IMAGING SCIENCES

Image Processing and Acquisition using Python

Ravishankar Chityala
University of Minnesota at Minneapolis
USA

Sridevi Pudipeddi
SriRav Scientific Solutions
Minneapolis, Minnesota, USA

CRC Press
Taylor & Francis Group
Boca Raton London New York

CRC Press is an imprint of the
Taylor & Francis Group an **informa** business

A CHAPMAN & HALL BOOK

MATLAB® is a trademark of The MathWorks, Inc. and is used with permission. The MathWorks does not warrant the accuracy of the text or exercises in this book. This book's use or discussion of MATLAB® software or related products does not constitute endorsement or sponsorship by The MathWorks of a particular pedagogical approach or particular use of the MATLAB® software.

CRC Press
Taylor & Francis Group
6000 Broken Sound Parkway NW, Suite 300
Boca Raton, FL 33487-2742

© 2014 by Taylor & Francis Group, LLC
CRC Press is an imprint of Taylor & Francis Group, an Informa business

No claim to original U.S. Government works

Printed on acid-free paper
Version Date: 20131206

International Standard Book Number-13: 978-1-4665-8375-7 (Hardback)

This book contains information obtained from authentic and highly regarded sources. Reasonable efforts have been made to publish reliable data and information, but the author and publisher cannot assume responsibility for the validity of all materials or the consequences of their use. The authors and publishers have attempted to trace the copyright holders of all material reproduced in this publication and apologize to copyright holders if permission to publish in this form has not been obtained. If any copyright material has not been acknowledged please write and let us know so we may rectify in any future reprint.

Except as permitted under U.S. Copyright Law, no part of this book may be reprinted, reproduced, transmitted, or utilized in any form by any electronic, mechanical, or other means, now known or hereafter invented, including photocopying, microfilming, and recording, or in any information storage or retrieval system, without written permission from the publishers.

For permission to photocopy or use material electronically from this work, please access www.copyright.com (http://www.copyright.com/) or contact the Copyright Clearance Center, Inc. (CCC), 222 Rosewood Drive, Danvers, MA 01923, 978-750-8400. CCC is a not-for-profit organization that provides licenses and registration for a variety of users. For organizations that have been granted a photocopy license by the CCC, a separate system of payment has been arranged.

Trademark Notice: Product or corporate names may be trademarks or registered trademarks, and are used only for identification and explanation without intent to infringe.

Library of Congress Cataloging-in-Publication Data

Chityala, Ravishankar, author.
  Image processing and acquisition using Python / Ravishankar Chityala, Sridevi Pudipeddi.
    pages cm. -- (Chapman & Hall/CRC mathematical and computational imaging sciences series)
    Summary: "If title belongs to a series, the exact series title that will appear in the book
is: Series number, exactly as it will appear in the book: Will the first-named author listed in Author Information appear as the first-named author for all other volumes of the series? "-- Provided by publisher.
    Includes bibliographical references and index.
    ISBN 978-1-4665-8375-7 (hardback)
    1. Image processing. 2. Python (Computer program language) I. Pudipeddi, Sridevi, author. II. Title.

TA1637.C486 2014
006.6'63--dc23   2013047010

Visit the Taylor & Francis Web site at
http://www.taylorandfrancis.com

and the CRC Press Web site at
http://www.crcpress.com

To our parents and siblings

Contents

List of Figures xvii

List of Tables xxiii

Foreword xxv

Preface xxvii

Introduction xxxi

About the Authors xxxiii

List of Symbols and Abbreviations xxxv

I Introduction to Images and Computing using Python   1

1 Introduction to Python   3
    1.1 Introduction 3
    1.2 What is Python? 4
    1.3 Python Environments 5
        1.3.1 Python Interpreter 6
        1.3.2 Enthought Python Distribution (EPD) 6
        1.3.3 PythonXY 7
    1.4 Running a Python Program 8
    1.5 Basic Python Statements and Data Types 8
        1.5.1 Data Structures 11
    1.6 Summary 19

1.7	Exercises		20

2 Computing using Python Modules — 23

- 2.1 Introduction — 23
- 2.2 Python Modules — 23
  - 2.2.1 Creating Modules — 24
  - 2.2.2 Loading Modules — 24
- 2.3 Numpy — 26
  - 2.3.1 Numpy Array or Matrices? — 30
- 2.4 Scipy — 31
- 2.5 Matplotlib — 32
- 2.6 Python Imaging Library — 33
- 2.7 Scikits — 33
- 2.8 Python OpenCV Module — 34
- 2.9 Summary — 34
- 2.10 Exercises — 35

3 Image and its Properties — 37

- 3.1 Introduction — 37
- 3.2 Image and its Properties — 38
  - 3.2.1 Bit Depth — 38
  - 3.2.2 Pixel and Voxel — 39
  - 3.2.3 Image Histogram — 41
  - 3.2.4 Window and Level — 42
  - 3.2.5 Connectivity: 4 or 8 Pixels — 43
- 3.3 Image Types — 44
  - 3.3.1 JPEG — 44
  - 3.3.2 TIFF — 44
  - 3.3.3 DICOM — 45
- 3.4 Data Structures for Image Analysis — 49
  - 3.4.1 Reading Images — 49
  - 3.4.2 Displaying Images — 50
  - 3.4.3 Writing Images — 50

3.5	Programming Paradigm	51
3.6	Summary	53
3.7	Exercises	53

II Image Processing using Python   55

4 Spatial Filters   57

- 4.1 Introduction 57
- 4.2 Filtering 58
  - 4.2.1 Mean Filter 60
  - 4.2.2 Median Filter 64
  - 4.2.3 Max Filter 66
  - 4.2.4 Min Filter 68
- 4.3 Edge Detection using Derivatives 69
  - 4.3.1 First Derivative Filters 71
  - 4.3.2 Second Derivative Filters 79
- 4.4 Summary 85
- 4.5 Exercises 86

5 Image Enhancement   89

- 5.1 Introduction 89
- 5.2 Pixel Transformation 89
- 5.3 Image Inverse 91
- 5.4 Power Law Transformation 92
- 5.5 Log Transformation 97
- 5.6 Histogram Equalization 99
- 5.7 Contrast Stretching 103
- 5.8 Summary 106
- 5.9 Exercises 107

6 Fourier Transform   109

- 6.1 Introduction 109
- 6.2 Definition of Fourier Transform 110
- 6.3 Two-Dimensional Fourier Transform 113

		6.3.1 Fast Fourier Transform using Python	115
	6.4	Convolution	118
		6.4.1 Convolution in Fourier Space	119
	6.5	Filtering in Frequency Domain	120
		6.5.1 Ideal Lowpass Filter	120
		6.5.2 Butterworth Lowpass Filter	123
		6.5.3 Gaussian Lowpass Filter	125
		6.5.4 Ideal Highpass Filter	127
		6.5.5 Butterworth Highpass Filter	130
		6.5.6 Gaussian Highpass Filter	132
		6.5.7 Bandpass Filter	134
	6.6	Summary	137
	6.7	Exercises	138

7 Segmentation — 139

	7.1	Introduction	139
	7.2	Histogram Based Segmentation	139
		7.2.1 Otsu's Method	141
		7.2.2 Renyi Entropy	144
		7.2.3 Adaptive Thresholding	149
	7.3	Region Based Segmentation	151
		7.3.1 Watershed Segmentation	153
	7.4	Segmentation Algorithm for Various Modalities	161
		7.4.1 Segmentation of Computed Tomography Image	161
		7.4.2 Segmentation of MRI Image	161
		7.4.3 Segmentation of Optical and Electron Microscope Image	162
	7.5	Summary	162
	7.6	Exercises	163

8 Morphological Operations — 165

	8.1	Introduction	165
	8.2	History	165

8.3	Dilation	166
8.4	Erosion	171
8.5	Grayscale Dilation and Erosion	175
8.6	Opening and Closing	176
8.7	Hit-or-Miss	179
8.8	Thickening and Thinning	184
	8.8.1 Skeletonization	185
8.9	Summary	186
8.10	Exercises	187

9 Image Measurements — 189

9.1	Introduction	189
9.2	Labeling	189
9.3	Hough Transform	194
	9.3.1 Hough Line	194
	9.3.2 Hough Circle	197
9.4	Template Matching	201
9.5	Summary	205
9.6	Exercises	205

III Image Acquisition — 207

10 X-Ray and Computed Tomography — 209

10.1	Introduction	209
10.2	History	209
10.3	X-Ray Generation	210
	10.3.1 X-Ray Tube Construction	210
	10.3.2 X-Ray Generation Process	212
10.4	Material Properties	216
	10.4.1 Attenuation	216
	10.4.2 Lambert Beer Law for Multiple Materials	218
10.5	X-Ray Detection	219
	10.5.1 Image Intensifier	220
	10.5.2 Multiple-Field II	221

	10.5.3 Flat Panel Detector (FPD)	223

- 10.6 X-Ray Imaging Modes 224
  - 10.6.1 Fluoroscopy 224
  - 10.6.2 Angiography 224
- 10.7 Computed Tomography (CT) 226
  - 10.7.1 Reconstruction 227
  - 10.7.2 Parallel Beam CT 227
  - 10.7.3 Central Slice Theorem 228
  - 10.7.4 Fan Beam CT 232
  - 10.7.5 Cone Beam CT 233
  - 10.7.6 Micro-CT 234
- 10.8 Hounsfield Unit (HU) 236
- 10.9 Artifacts 237
  - 10.9.1 Geometric Misalignment Artifacts 238
  - 10.9.2 Scatter 238
  - 10.9.3 Offset and Gain Correction 240
  - 10.9.4 Beam Hardening 241
  - 10.9.5 Metal Artifacts 242
- 10.10 Summary 243
- 10.11 Exercises 244

11 Magnetic Resonance Imaging   247

- 11.1 Introduction 247
- 11.2 Laws Governing NMR and MRI 248
  - 11.2.1 Faraday's Law 248
  - 11.2.2 Larmor Frequency 249
  - 11.2.3 Bloch Equation 250
- 11.3 Material Properties 251
  - 11.3.1 Gyromagnetic Ratio 251
  - 11.3.2 Proton Density 252
  - 11.3.3 T_1 and T_2 Relaxation Times 253
- 11.4 NMR Signal Detection 255
- 11.5 MRI Signal Detection or MRI Imaging 256

		11.5.1 Slice Selection	258
		11.5.2 Phase Encoding	258
		11.5.3 Frequency Encoding	259
	11.6	MRI Construction	259
		11.6.1 Main Magnet	259
		11.6.2 Gradient Magnet	260
		11.6.3 RF Coils	261
		11.6.4 K-Space Imaging	262
	11.7	T_1, T_2 and Proton Density Image	263
	11.8	MRI Modes or Pulse Sequence	265
		11.8.1 Spin Echo Imaging	265
		11.8.2 Inversion Recovery	266
		11.8.3 Gradient Echo Imaging	267
	11.9	MRI Artifacts	268
		11.9.1 Motion Artifact	269
		11.9.2 Metal Artifact	271
		11.9.3 Inhomogeneity Artifact	271
		11.9.4 Partial Volume Artifact	272
	11.10	Summary	272
	11.11	Exercises	273

12 Light Microscopes   275

	12.1	Introduction	275
	12.2	Physical Principles	276
		12.2.1 Geometric Optics	276
		12.2.2 Numerical Aperture	277
		12.2.3 Diffraction Limit	278
		12.2.4 Objective Lens	280
		12.2.5 Point Spread Function (PSF)	281
		12.2.6 Wide-Field Microscopes	282
	12.3	Construction of a Wide-Field Microscope	282
	12.4	Epi-Illumination	284
	12.5	Fluorescence Microscope	284

	12.5.1	Theory	284
	12.5.2	Properties of Fluorochromes	285
	12.5.3	Filters	287
12.6	Confocal Microscopes	288
12.7	Nipkow Disk Microscopes	289
12.8	Confocal or Wide-Field?	291
12.9	Summary	292
12.10	Exercises	293

13 Electron Microscopes   295

13.1	Introduction	295
13.2	Physical Principles	296
	13.2.1	Electron Beam	297
	13.2.2	Interaction of Electron with Matter	298
	13.2.3	Interaction of Electrons in TEM	299
	13.2.4	Interaction of Electrons in SEM	300
13.3	Construction of EM	301
	13.3.1	Electron Gun	301
	13.3.2	Electromagnetic Lens	303
	13.3.3	Detectors	304
13.4	Specimen Preparations	306
13.5	Construction of TEM	307
13.6	Construction of SEM	308
13.7	Summary	309
13.8	Exercises	311

A Installing Python Distributions   313

A.1	Windows	313
	A.1.1	PythonXY	313
	A.1.2	Enthought Python Distribution	316
	A.1.3	Updating or Installing New Modules	316
A.2	Mac or Linux	318
	A.2.1	Enthought Python Distribution	318

| | | A.2.2 Installing New Modules . . . . | 318 |

B Parallel Programming Using MPI4Py — 323

- B.1 Introduction to MPI 323
- B.2 Need for MPI in Python Image Processing 324
- B.3 Introduction to MPI4Py 325
- B.4 Communicator 326
- B.5 Communication 327
  - B.5.1 Point-to-Point Communication 327
  - B.5.2 Collective Communication 329
- B.6 Calculating the Value of PI 331

C Introduction to ImageJ — 333

- C.1 Introduction 333
- C.2 ImageJ Primer 334

D MATLAB® and Numpy Functions — 337

- D.1 Introduction 337

Bibliography — 341

Index — 351

List of Figures

1.1	PythonXY command prompt without the IDE.	7
2.1	Example of a plot generated using matplotlib.	32
3.1	Image processing work flow.	37
3.2	Width and height of pixel in physical space.	40
3.3	An example of volume rendering.	41
3.4	An example of a histogram.	42
3.5	Window and level.	43
3.6	An example of 4 and 8 pixel connectivity.	43
4.1	An example of different padding options.	61
4.2	Example of mean filter.	64
4.3	Example of median filter.	67
4.4	Example of max filter.	68
4.5	Example of min filter.	69
4.6	An example of zero-crossing.	71
4.7	Example for Sobel and Prewitt.	75
4.8	Output from vertical, horizontal and regular Sobel and Prewitt filters.	77
4.9	Example of Canny filter.	80
4.10	Example of the Laplacian filter.	82
4.11	Another example of Laplacian filter.	83
4.12	Example of LoG.	86
5.1	Illustration of transformation $T(x) = x^2$.	90
5.2	Example of transformation $T(x) = x + 50$.	91

5.3	Example of image inverse transformation.	93
5.4	Graph of power law transformation for different γ.	94
5.5	An example of power law transformation.	96
5.6	Graph of log and inverse log transformations.	98
5.7	Example of log transformation.	99
5.8	An example of a 5-by-5 image.	101
5.9	Probabilities, CDF, histogram equalization transformation.	102
5.10	Example of histogram equalization.	104
5.11	An example of contrast stretching where the pixel value range is significantly different from $[0, 255]$.	106
5.12	An example of contrast stretching where the input pixel value range is same as $[0, 255]$.	106
6.1	An example of 2D Fast Fourier transform.	117
6.2	An example of lowpass filters. The input image and all the output images are displayed in spatial domain.	128
6.3	An example of highpass filters. The input image and all the output images are displayed in spatial domain.	134
6.4	An example of IBPF. The input and the output are displayed in spatial domain.	137
7.1	Threshold divides the pixels into foreground and background.	140
7.2	An example of Otsu's method.	143
7.3	Another example of Otsu's method.	144
7.4	An example of Renyi entropy.	148
7.5	An example of thresholding with adaptive vs. Otsu's.	151
7.6	An example of an image for region-based segmentation.	152
7.7	An example of watershed segmentation.	160
8.1	An example of binary dilation.	167
8.2	An example of binary dilation.	171
8.3	An example of binary erosion.	172

List of Figures

8.4	An example of binary erosion.	175
8.5	An example of binary opening with 5 iterations.	178
8.6	An example of binary closing with 5 iterations.	179
8.7	An example of hit-or-miss transformation.	181
8.8	An example of hit-or-miss transformation on a binary image.	183
8.9	An example of skeletonization.	187
9.1	An example of regionprops.	194
9.2	An example of Hough line transform.	197
9.3	An example of Hough circle transform.	201
9.4	An example of template matching.	204
10.1	Components of an x-ray tube.	211
10.2	X-ray spectrum illustrating characteristic and Bremsstrahlung spectrum.	213
10.3	Production of Bremsstrahlung or braking spectrum.	214
10.4	Production of characteristic radiation.	215
10.5	Lambert Beer law for monochromatic radiation and for a single material.	217
10.6	Lambert Beer law for multiple materials.	219
10.7	Ionization detector.	220
10.8	Components of an image intensifier.	222
10.9	Flat panel detector schematic.	223
10.10	Fluoroscopy machine.	225
10.11	Parallel beam geometry.	228
10.12	Central slice theorem.	229
10.13	Fan beam geometry.	232
10.14	Axial CT slice.	233
10.15	Montage of all the CT slices of the human kidney region.	233
10.16	3D object created using the axial slices shown in the montage. The 3D object in green is superimposed on the slice information for clarity.	234

10.17	Cone beam geometry.	235
10.18	Parameters defining a cone beam system.	239
10.19	Scatter radiation.	240
10.20	Effect of metal artifact.	243
11.1	Illustration of Faraday's law.	249
11.2	Precessing of nucleus in a magnetic field.	250
11.3	Bloch equation as a 3D plot.	251
11.4	T_1 magnetization.	254
11.5	Plot of T_1 magnetization.	254
11.6	Plot of T_2 de-magnetization.	255
11.7	Net magnetization and effect of RF pulse.	257
11.8	Free induction decay.	257
11.9	Slice selection gradient.	258
11.10	Closed magnet MRI machine.	260
11.11	Open magnet MRI machine.	261
11.12	Head coil.	262
11.13	k-space image.	263
11.14	k-space reconstruction of MRI images.	264
11.15	T_1, T_2 and proton density image.	265
11.16	Spin echo pulse sequence.	266
11.17	Inversion recovery pulse sequence.	267
11.18	Gradient echo pulse sequence.	268
11.19	Effect of motion artifact on MRI reconstruction.	270
11.20	Metal artifact formation.	271
12.1	Light microscope.	277
12.2	Schematic of the light microscope.	278
12.3	Markings on the objective lens.	279
12.4	Rayleigh Criterion.	280
12.5	Jablonski diagram.	285
12.6	Nipkow disk design.	290
12.7	Nipkow disk setup.	290
12.8	Photograph of Nipkow disk microscope.	291

List of Figures

13.1	Intensity distributions.	299
13.2	Thermionic gun.	302
13.3	Field emission gun.	303
13.4	Electromagnetic lens.	304
13.5	Everhart-Thornley secondary electron detector.	305
13.6	Back-scattered electron detector.	306
13.7	Comparison of optical microscope, TEM and SEM.	308
13.8	TEM slice and its iso-surface rendering.	308
13.9	An SEM machine.	309
13.10	BSE image obtained using an SEM.	310
A.1	Specifying the type of install.	314
A.2	The Windows menu item to start PythonXY under Spyder.	315
A.3	The Spyder interface.	315
A.4	Specifying a Python distribution for installation.	316
A.5	Installation of skimage module.	317
A.6	Specifying the interpreter version and location.	317
A.7	Installing Enthought Python distribution on Mac.	319
A.8	Loading Enthought Python distribution on Mac and skimage module.	319
A.9	Installing cython module using easy_install. This module is required to use skimage module.	320
A.10	Installing skimage module using easy_install.	320
A.11	Loading skimage module.	321
A.12	Steps for installing pydicom on Windows.	321
C.1	ImageJ main screen	334
C.2	ImageJ with an MRI image.	335
C.3	Adjusting window or level on an MRI image.	335
C.4	Performing median filter.	336
C.5	Obtaining histogram of the image.	336

List of Tables

4.1	A 3-by-3 filter.	58
4.2	A 3-by-3 sub-image.	59
4.3	Sobel masks for horizontal and vertical edges.	72
4.4	A 3-by-3 subimage.	72
4.5	Output after multiplying the sub-image with Sobel masks.	73
4.6	Prewitt masks for horizontal and vertical edges.	73
4.7	Sobel masks for diagonal edges.	75
4.8	Prewitt masks for diagonal edges.	76
4.9	Laplacian masks.	80
4.10	Laplacian of Gaussian mask	83
8.1	Hit-or-miss structuring element	180
8.2	Variation of all structuring elements used to find corners.	180
10.1	Relationship between kVp and HVL.	218
11.1	An abbreviated list of the nuclei of interest to NMR and MRI imaging and their gyromagnetic ratios.	252
11.2	List of biological materials and their proton or spin density.	253
11.3	List of biological materials and their T_1 and T_2 values for field strength of 1.0 T.	255
11.4	TR and TE settings for various weighted images.	268

12.1 List of the commonly used media and their refractive indexes. 281

12.2 List of the fluorophores of interest to fluorescence imaging. 286

Foreword

I first met one of the authors, Dr. Ravishankar (Ravi) Chityala, in 2006 when he was a PhD student at the Toshiba Stroke Research Center, SUNY-Buffalo. Ravi's PhD work in medical imaging was fruitful and influential, and I have been following his post-PhD career ever since. In reading this book, I was impressed by the fact that, despite Ravi's current focus on computing and visualization, his knowledge of medical imaging has only deepened and expanded, which has enabled him, along with his co-author, Dr. Sridevi Pudipeddi, to write a very competent treatment of the subject of medical imaging. Thus, it is a pleasure for me to write a foreword to this very good book.

This is a book that every imaging scientist should have on his or her desk because image acquisition and processing is becoming a standard method for qualifying and quantifying experimental measurements. Moreover, I believe students and researchers need a course or a book to learn both image acquisition and image processing using a single source, and this book, as a well-rounded introduction to both topics, serves that purpose very well. The topics treated are complex, but the authors have done a great job of covering the most commonly used image acquisition modalities, such as x-ray and computed tomography, magnetic resonance imaging, and microscopes, concisely and effectively, providing a handy compendium of the most useful information.

As Confucius said, "I see and I remember, I do and I understand;" this book aims to provide hands-on learning that enables the reader to understand the concepts explained in the book by means of applying the various examples written in the Python code. But do not be discouraged if you have never used Python or any other script language since learning it is very straightforward. As a long-time Perl user, I had no problem installing Python and trying several useful examples from the book. Most of the equations provided in the book are accompanied by codes that can be quickly run and modified for the reader to test new ideas and apply to his or her own research.

Being a medical imaging scientist myself, I really enjoyed reading the sections on x-ray, computed tomography and magnetic resonance imaging. The authors provide a well-balanced introduction to these modalities and cover all the important aspects of image acquisition, as well as image reconstruction and artifacts correction. The authors also provide a large number of references to other books and papers for readers interested in learning more details.

In summary, the strengths of the book are:

1. It teaches image processing using Python, one of the easiest and most powerful programming languages

2. It covers commonly used image acquisition and processing techniques

3. It cements readers' understanding with numerous clear examples.

Alexander Zamyatin
Distinguished Scientist
Toshiba Medical Research Institute USA, Inc.
Vernon Hills, Illinois

Preface

Image acquisition and processing have become a standard method for qualifying and quantifying experimental measurements in various Science, Technology, Engineering, and Mathematics (STEM) disciplines. Discoveries have been made possible in medical sciences by advances in diagnostic imaging such as x-ray based computed tomography (CT) and magnetic resonance imaging (MRI). Biological and cellular functions have been revealed with new imaging techniques in light based microscopy. Advancements in material sciences have been aided by electron microscopy. All these examples and many more require knowledge of both the physical methods of obtaining images and the analytical processing methods to understand the science behind the images. Imaging technology continues to advance with new modalities and methods available to students and researchers in STEM disciplines. Thus, a course in image acquisition and processing has broad appeal across the STEM disciplines and is useful for transforming undergraduate and graduate curriculum to better prepare students for their future.

This book covers both image acquisition and image processing. Existing books discuss either image acquisition or image processing, leaving a student to rely on two different books containing different notations and structures to obtain a complete picture. Integration of the two is left to the readers.

During the authors' combined experiences in image processing, we have learned the need for image processing education. We hope this book will provide sufficient background material in both image acquisition and processing.

Audience

The book is intended primarily for advanced undergraduate and graduate students in applied mathematics, scientific computing, medical imaging, cell biology, bioengineering, computer vision, computer science, engineering and related fields, as well as to engineers, professionals from academia, and the industry. The book can be used as a textbook for an advanced undergraduate or graduate course, a summer seminar course, or can be used for self-learning. It serves as a self-contained handbook and provides an overview of the relevant image acquisition techniques and corresponding image processing. The book also contains practice exercises and tips that students can use to remember key information.

Acknowledgments

We are extremely thankful to students, colleagues, and friends who gave valuable input during the process of writing this book. We are thankful to the Minnesota Supercomputing Institute (MSI) at the University of Minnesota. At MSI, Ravi Chityala had discussions with students, staff and faculty on image processing. These discussions helped him recognize the need for a textbook that combines both image processing and acquisition.

We want to specially thank Dr. Nicholas Labello, University of Chicago; Dr. Wei Zhang, University of Minnesota; Dr. Guillermo Marques, University Imaging Center, University of Minnesota; Dr. Greg Metzger, University of Minnesota; Mr. William Hellriegel, University of Minnesota; Dr. Andrew Gustafson, University of Minnesota; Mr. Abhijeet More, Amazon; Mr. Arun Balaji; and Mr. Karthik Bharathwaj for proofreading the manuscript and for providing feedback.

We thank Carl Zeiss Microscopy; Visible Human Project; Siemens AG; Dr. Uma Valeti, University of Minnesota; Dr. Susanta Hui, University of Minnesota; Dr. Robert Jones, University of Minnesota; Dr. Wei Zhang, University of Minnesota; Mr. Karthik Bharathwaj for providing us with images that were used in this book.

We also thank our editor Sunil Nair and editorial assistant Sarah Gelson; project coordinator Laurie Schlags; project editor Amy Rodriguez at Taylor and Francis/CRC Press for helping us during the proofreading and publication process.

MATLAB® is a registered trademark of The MathWorks, Inc. For product information, please contact:
The MathWorks, Inc.
3 Apple Hill Drive
Natick, MA 01760-2098 USA
Tel: 508-647-7000
Fax: 508-647-7001
E-mail: info@mathworks.com
Web: www.mathworks.com

Introduction

This book is meant for upper level undergraduates, graduate students and researchers in various disciplines in STEM. The book covers both image acquisition and image processing. The knowledge of image acquisition will help readers to perform experiments more effectively and cost efficiently. The knowledge of image processing will help the reader to analyze and measure efficiently. The concepts of image processing will become ingrained using examples written using Python, long recognized as one of the easiest languages for non-programmers to learn.

Python is a good choice for teaching image processing because:

1. It is freely available and open source. Since it is free software, all students will have access to it without any restriction

2. It provides pre-packed installations available for all major platforms at no cost

3. It is the high-level language of choice for scientists and engineers

4. It is recognized as perhaps the easiest language to learn for non-programmers

Due to new developments in imaging technology as well as the scientific need for higher resolution images, the image data sets are getting larger every year. Such large data sets can be analyzed quickly using a large number of computers. Closed source software like MATLAB® cannot be scaled to a large number of computers as the licensing cost is high. On the other hand, Python, being free and open-source software, can be scaled to thousands of computers at no cost. For these reasons, we strongly believe the future need for image processing for all students can be met effectively using Python.

The book consists of three parts: Python programming, image processing, and image acquisition. Each of these parts consists of multiple chapters. The parts are self-contained. Hence, a user well versed in Python programming can skip Part I and read only Parts II and III. Each chapter contains many examples, detailed derivations, and working Python examples of the techniques discussed within. The chapters are also interspersed with practical tips on image acquisition and processing. The end of every chapter contains a summary of the important points discussed and a list of exercise problems to cement the reader's understanding.

Part I consists of introduction to Python, Python modules, reading and writing images using Python, and an introduction to images. Readers can skip or skim this part if they are already familiar with the material. This part is a refresher and readers will be directed to other resources as applicable.

In Part II we discuss the basics of image processing. The various chapters discuss pre/post processing using filters, segmentation, morphological operations and measurements.

In Part III we discuss image acquisition using various modalities like x-ray, CT, MRI, light microscopy and electron microscopy. These modalities cover most of the common image acquisition methods used currently by researchers in academia and industry.

Details about exercises

The Python programming and image processing parts of the book contain exercises that test the reader's skills in Python programming, image processing, and integration of the two. Solutions to odd-numbered problems, example programs and images are available at http://www.crcpress.com/product/isbn/9781466583757.

About the Authors

Ravishankar Chityala, Ph.D. is an image processing consultant at the Minnesota Supercomputing Institute of the University of Minnesota, with more than eleven years experience in image processing. As an image processing consultant, Dr. Chityala has worked with faculty, students and staff from various departments in the scientific, engineering and medical fields at the University of Minnesota, and his interaction with students has made him aware of their need for greater understanding of and ability to work with image processing and acquisition. Dr. Chityala contributed to the writing of *Handbook of Physics in Medicine and Biology* (CRC Press, Boca Raton, 2009, Robert Splinter).

Sridevi Pudipeddi, Ph.D. is an image processing consultant at SriRav Scientific Solutions, Minneapolis, Minnesota, with eleven years experience teaching undergraduate courses. Dr. Pudipeddi's research interests are in applied mathematics and image and text processing. Python's simple syntax and its vast image processing capabilities, along with the need to understand and quantify important experimental information through image acquisition, have inspired her to co-author this book.

List of Symbols and Abbreviations

$\sum$	summation
θ	angle
$\lvert x \rvert$	absolute value of x
e	2.718281
$*$	convolution
$\log$	logarithm base 10
$\ominus$	morphological erosion
$\oplus$	morphological dilation
$\circ$	morphological open
$\bullet$	morphological close
$\cup$	union
λ	wavelength
E	energy
h	Planck's constant
c	speed of light
μ	attentuation coefficient
γ	gyromagnetic ratio
NA	numerical aperture
ν	frequency
dx	differential
∇	gradient
$\frac{\partial}{\partial x}$	derivative along x-axis
$\nabla^2 = \Delta$	Laplacian
$\int$	integration
CDF	cumulative distribution function
CT	computed tomography
DICOM	digital imaging and communication in medicine
JPEG	joint photographic experts group
MRI	magnetic resonance imaging
PET	positron emission tomography
PNG	portable network graphics
PSF	point spread function
RGB	red, green, blue channels
TIFF	tagged image file format

Part I

Introduction to Images and Computing using Python

Chapter 1

Introduction to Python

1.1 Introduction

Before we begin discussion on image acquisition and processing using Python, we will provide an overview of the various aspects of Python. This chapter focuses on some of the basic materials covered by many other books [2], [36], [54], [103]. If you are already familiar with Python and are currently using it then you can skip this chapter.

We begin with an introduction to Python. We discuss the installation of Python with all the modules. The details of the installation process are available in Appendix A. To simplify the installation, we will discuss the Python distributions Enthought Python Distribution (EPD) and PythonXY. If you are an advanced programmer, you can install the various modules like scipy, numpy, Python imaging library etc. in the basic Python distribution. Once the installation has been completed, we can begin exploring the various features of Python. We will quickly review the various data structures such as list, dictionary, and tuples and statements such as for-loop, if-else, iterators and list comprehension.

1.2 What is Python?

Python is a popular high-level programming language. It can handle various programming tasks such as numerical computation, web development, database programming, network programming, parallel processing, etc. Python is popular for various reasons including:

1. It is free.

2. It is available on all the popular operating systems such as Windows, Mac or Linux.

3. It is an interpreted language. Hence, programmers can test portions of code on the command line before incorporating it into their program. There is no need for compiling or linking.

4. It gives the ability to program faster.

5. It is syntactically simpler than C/C++/Fortran. Hence it is highly readable and easier to debug.

6. It comes with various modules that are standard or can be installed to an existing Python installation. These modules can perform various tasks like reading and writing various files, scientific computation, visualization of data etc.

7. Programs written in Python can be imported to various OS or platforms with little or no change.

8. It is a dynamically typed language. Hence the data type of variables does not have to be declared prior to their use.

9. It has a dedicated developer and user community and is kept up to date.

Although Python has many advantages that have made it one of the most popular interpreted languages, it has a couple of drawbacks that are discussed below:

1. Since its focus is on the ability to program faster, the speed of execution suffers. A Python program might be 10 times or more slower than an equivalent C program but it will contain fewer lines of code and can be programmed to handle multiple data types easily. This drawback in the Python code can be overcome by converting the computationally intensive portions of the code to C/C++ or by the appropriate use of data structure.

2. Indentation of the code is not optional. This makes the code readable. However, a code with multiple loops and other constructs will be indented to the right, making it difficult to read the code. Python provides some tools such as list processing, dictionary and set to reduce this complexity.

1.3 Python Environments

There are several Python environments from which to choose. Some operating systems like Mac, Linux, Unix etc. have a built-in interpreter. The interpreter may contain all modules but is not turn-key ready for scientific processing. Specialized distributions have been created and sold to the scientific community, pre-built with various Python scientific modules. The users do not have to install the individual scientific modules. If a particular module that is of interest is not available in the distribution, it can be installed using the existing distribution. Some of the most popular distributions are Enthought Python Distribution and PythonXY. The instructions for installing these distrubtions are available in Appendix A, Installing Python Distributions.

1.3.1 Python Interpreter

If you are using Mac or Linux, you are lucky, as Python interpreters are built in to the operating system. Python can be started by simply typing **python** in the terminal window. However, in Windows, a programming environment such as IDLE needs to be installed. When the interpreter is started, a command prompt (>>>) appears. Python commands can be entered at the prompt for processing. In Mac, when the built-in Python interpreter is started, an output similar to the one shown below appears:

```
macbook:\~ chityala$ /usr/bin/python
Python 2.5.1 (r251:54863, May  5 2011, 18:37:34)
[GCC 4.0.1 (Apple Inc. build 5465)] on darwin
Type "help", "copyright", "credits" or "license"
for more information.
>>>
```

1.3.2 Enthought Python Distribution (EPD)

The Enthought Python Distribution [99] provides programmers with close to 100 of the most popular scientific Python modules like scientific computation, linear algebra, symbolic computing, image processing, signal processing, visualization, integration of C/C++ programs to Python etc. It is distributed and maintained by Enthought Scientific Computing Solutions. It is available for free for academics and is available for a price to all others. In addition to the various modules built in to EPD, programmers can install other modules using virtualenv [3], without affecting the main distribution.

In Mac, when the EPD Python interpreter is started, an output similar to the one shown below appears:

```
macbook:~ chityala$ python
Enthought Python Distribution -- www.enthought.com
```

```
Version: 7.2-2 (64-bit)
Python 2.7.2 |EPD 7.2-2(64-bit)|
        (default, Sep 7 2011, 16:31:15)
[GCC 4.0.1 (Apple Inc. build 5493)] on darwin
Type "packages", "demo" or "enthought"
for more information.
>>>
```

1.3.3 PythonXY

PythonXY [78] is a free scientific Python distribution. It is pre-built with many scientific modules similar to Enthought Python distribution. These modules can be used for developing Python based applications that need numerical computing and visualization. It is also shipped with an integrated development environment called Spyder.

Spyder is designed to have the same look and feel of the MATLAB® development environment, and ease the task of migrating from MATLAB to Python. It also has a built-in documentation of the various functions and classes available in its modules.

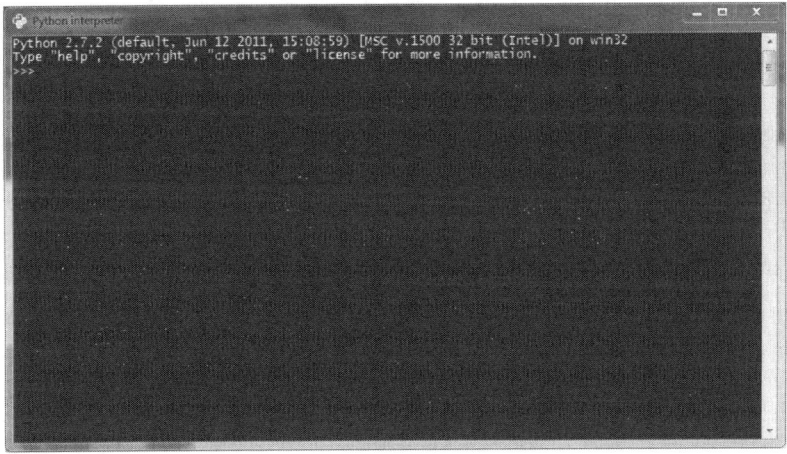

FIGURE 1.1: PythonXY command prompt without the IDE.

1.4 Running a Python Program

Using any interpreter, you can run your program using the command at the Linux or Mac command prompt.

```
>> python firstprog.py
```

The >> is the terminal prompt and >>> represents the Python prompt.

In the case of Windows, you can run the program by double-clicking the .py file or by typing the name of the program in the Windows command prompt. The best approach to running Python programs under Windows is to use an Integrated Development Environment like IDLE, Spyder or EPD.

1.5 Basic Python Statements and Data Types

Indentation

The general practice is to add a tab or 4 spaces at the start of every line. Since indentation is part of the language to separate logical code sections, users do not need to use any curly braces. The start and end of a block can be easily identified by following the indentation. There is a significant disadvantage to indentation. A code containing multiple for-loops and if-statements will be indented farther to the right making the code unreadable. This problem can be alleviated by reducing the number of for-loops and if-statements. This not only makes the code readable but also reduces computational time. This can be achieved by programming using more data structures like lists, dictionary, and set and by using operators such as iterators.

Comments

Comments are an important part of any programming language. In Python, a comment is denoted by a hash mark # at the beginning of a line. Multiple lines can be commented by using triple quoted strings (''') at the beginning and at the end of the block.

```
# This is a single line comment

'''
This is
a multiline
comment
'''

# Comments are a good way to explain the code.
```

Variables

Python is a dynamic language and hence you do not need to specify the variable type as in C/C++. Variables can be imagined as containers of values. The values can be an integer, float, string, etc.

```
>>> a = 1
>>> a = 10.0
>>> a = 'hello'
```

In the above example the integer value of 1, float value of 10.0 and a string value of `hello` for all cases are stored in the same variable. However, only the last assigned value is the current value for a.

Operators

Python supports all the common arithmetic operators such as $+, -, *, /$. It also supports the common comparison operators such as $>, <, ==, !=, >=, <=$ etc. In addition, through various modules Python provides many operators for performing trigonometric, mathematical, geometric operations etc.

Loops

The most common looping construct in Python is the `for-loop` statement, which allows iterating through the collection of objects. Here is an example:

```
>>> for i in range(1,5)
...     print i
```

In the above example the output of the `for-loop` is the numbers from 1 to 5. The range function allows us to create values starting from 1 and ending with 5. Such a concept is similar to the `for-loop` normally found in C/C++ or most programming languages.

The real power of `for-loop` lies in its ability to iterate through other Python objects such as lists, dictionaries, sets, strings etc. We will discuss these Python objects in more detail subsequently.

```
>>> a = ['python','scipy',2.7]
>>> for i in a:
...        print i
```

In the program above, the `for-loop` iterates through each element of the list and prints it.

In the next program, the content of a dictionary is printed using the `for-loop`. A dictionary with two keys `lang` and `ver` is defined. Then, using the `for-loop` the various keys are iterated and the corresponding values are printed.

```
>>> a = {
         'lang':'python'
         'ver': '2.7.1'
        }
>>> for keys in a:
...      print a[key]
```

The discussion about using `for-loop` for iterating through the various lines in a text file such as comma separated value file is postponed to a later section.

if-else statement

The `if-else` is a popular conditional statement in all programming languages including Python. An example of `if-elif-else` statement is shown below.

```
if a<10:
    print 'a is less than 10'
elif a<20:
    print 'a is between 10 and 20'
else:
    print 'a is greater than 20'
```

The `if-else` statement conditionals do not necessarily have to use the conditional operators such as $<, >, ==$ etc.

For example, the following `if` statement is legal in Python. This `if` statement checks for the condition that the list `d` is not empty.

```
>>> d = [ ]
>>> if d:
...     print 'd is not empty'
... else:
...     print 'd is empty'
d is empty
```

1.5.1 Data Structures

The real power of Python lies in the liberal usage of its data structure. The common criticism of Python is that it is slow compared to C/C++. This is especially true if multiple for-loops are used in programming Python. Since Python is a dynamically typed language and since you do not declare the type and size of the variable, Python has to dynamically allocate space for variables that increase in size at runtime. Hence for-loops are not the most efficient way for programming Python. The alternate is to use data structures such as lists, tuples, dictionary and sets. We describe each of these data structures in this section.

Lists

Lists are similar to arrays in C/C++. But, unlike arrays in C/C++, lists in Python can hold objects of any type such as int, float, string and including another list. Lists can also have objects of different type. Lists are mutable, as their size can be changed by adding or removing elements. The following examples will help show the power and flexibility of lists.

```
>>> a = ['python','scipy',2.7]
>>> a.pop(-1)
>>> print a
a = ['python','scipy']
>>> a.insert(2,'numpy')
>>> print a[0]
scipy
>>> print a[-1]
numpy
>>> print a[0:2]
['python','scipy']
```

In the first line, a new list is created. This list contains two strings and one floating-point number. In the second line, we use pop function to remove the last element. So, when we print the content of the last element in line 3, we obtain a list containing only two elements instead of the original three. In line 5, we insert a new element, "numpy" at the end of the list or at position 2. Finally, in the next two commands we print the value of the list in position 0 and the last position indicated by using "−1" as the index. This indicates that one can operate on the list both using functions such as pop, insert or remove, slicing the list as we would slice a numerical array. In the last command, we introduce slicing and obtain a new list that contains only the first two values of the list.

A list may contain another list. Here is an example. We will consider the case of a list containing four numbers and arranged to look like and operate like a matrix.

```
>>> a = [[1,2],[3,4]]
>>> print a[0]
[1,2]
>>> print a[1]
[3,4]
>>> print a[0][0]
1
```

In line 1, we define a list of the list. The values [1, 2] are in the first list and the values [3, 4] are in the second list. The two lists have been combined to form a 2D list. In the second line, we print the value of the first element of the list. Note that this prints the first row or the first list and not just the first cell. In the fourth line, we print the value of the second row or the second list. To obtain the value of the first element in the first list we need to slice the list as given in line 6. As you can see, indexing the various elements of the list is as simple as calling the location of the list. Although in this example, we define a 2D list, one can create three, four or any higher order lists.

Although the list elements can be individually operated, the power of Python is in its ability to operate on the entire list at once using list functions and list comprehensions.

List functions

Let us consider the list that we created in the previous section. We can sort the list using the sort function as shown in line 2. The sort function does not return a list; instead, it modifies the current list. Hence the existing list will contain the elements in a sorted order. If a list contains both numbers and strings, Python sorts the numerical values first and then sorts the strings in alphabetical order.

```
>>> a = ['python','scipy','numpy']
>>> a.sort()
>>> a
['numpy','python','scipy']
```

List comprehensions

A list comprehension allows building a list from another list. Let us consider the case where we need to generate a list of squares of numbers from 0 to 9. We will begin by generating a list of numbers from 0 to 9. Then we will determine the square of each element.

In line number 1, a list is created containing values from 0 to 9 using the function "range" and print command is given in line 2. In line number 4, list comprehension is performed by taking each element in a and multiplying by itself. The result of the list comprehension is shown in line 5. The same operation can be performed by using lines 6 to 8 but the list comprehension approach is compact in syntax as it eliminates two lines of code, one level of indentation and a for-loop. It is also much faster when applied to a large list.

For a new Python programmer, the list comprehension might seem daunting. The best way to understand and read a list comprehension is by imagining that you will first operate on the for-loop and then begin reading the left part of the list comprehension. In addition to applying for-loop using list comprehension, you can also apply logical operations like if-else.

```
>>> a = range(10)
>>> print a
[0, 1, 2, 3, 4, 5, 6, 7, 8, 9]
>>> b = [x*x for x in a]
[0, 1, 4, 9, 16, 25, 36, 49, 64, 81]
>>> b = []
```

```
>>> for x in a:
        b.append(x*x)
>>> print b
[0, 1, 4, 9, 16, 25, 36, 49, 64, 81]
```

Tuples

Tuples are similar to lists except that they are not mutable, i.e. the length and the content of the tuple cannot be changed at runtime. Syntactically, the list uses [ ] while tuples use ( ). Similar to lists, tuples may contain any data type including other tuples. Here are a few examples:

```
>>> a = (1,2,3,4)
>>> print a
(1,2,3,4)
>>> b = (3,)
>>> c = ((1,2),(3,4))
```

In line 1, we define a tuple containing four elements. In line 4, we define a tuple containing only one element. Although the tuple contains only one element, we need to add the trailing comma, so that Python understands it as a tuple. Failure to add a comma at the end of this tuple will result in the value 3 being treated as an integer and not a tuple. In line 5, we create a tuple inside another tuple.

Sets

A set is an unordered collection of objects. A set can contain objects of any data type supported by Python. To create a set, we need to use the function set. Here are some examples:

```
>>> s1 = set([1,2,3,4])
>>> s2 = set((1,1,3,4))
>>> print s2
set([1,3,4])
```

In line 1, we create a set from a list containing four values. In line 2, we create a set containing a tuple. The elements of a set need to be unique. Hence when the content of s2 is printed, we notice that the duplicates have been eliminated. Sets in Python can be operated using many of the common mathematical operations on sets such as union, intersection, set difference, symmetric difference etc.

Since sets do not store repeating values and since we can convert lists and tuples to sets easily, they can be used to perform useful operations faster which otherwise would involve multiple loops and conditional statements. For example, a list containing only unique values can be obtained by converting the list to a set and back to a list. Here is an example:

```
>>> a = [1,2,3,4,3,5]
>>> b = set(a)
>>> print b
set([1,2,3,4,5])
>>> c = list(b)
>>> print c
[1,2,3,4,5]
```

In line 1, we create a list containing six values with one duplicate. We convert the list into a set by using the set() function. During this process, the duplicate value 3, has been eliminated. We can then convert the set back to list using the list() function.

Dictionaries

Dictionaries store a key-value pair. A dictionary is created by enclosing a key-value pair inside { }.

```
>>> a = {
            'lang':'python'
            'ver': '2.7.1'
         }
```

Any member of the dictionary can be accessed using

```
>>> print a['lang']
python
```

To add a new key,

```
>>> a['creator'] = 'Guido Von Rossum'
```

To remove a key, use the del method as shown below

```
>>> del a['ver']
```

In the example above, we added a new key called creator and stored the the string, "Guido Von Rossum."

In certain instances, the dictionary membership needs to be tested using the has_key() method. To obtain a list of all the dictionary keys, use the keys() method.

File handling

This book is on image processing; however, it is important to understand and be able to include in your code, reading and writing of text files so that the results of computation or the input parameters can be read from external sources. Python provides the ability to read and write files. It also has functions and methods for reading specialized formats such as csv, Microsoft Excel (xls) format etc. We will look into each method in this section.

```
>>> fo = open('myfile.csv')
>>> for i in fo.readlines():
...     print i
>>> fo.close()
Python, 2.7.1

Django, 1.4.1

Apache, 2.4
```

The first line opens a file and returns a new file object which is stored in the variable "fo." The method `readlines` in line 2, reads all the lines of input. The for-loop then iterates over each of those lines, and prints. The file is finally closed using the close method.

The output of the print command is a string. Hence, string manipulation using methods like split, strip etc. needs to be applied in order to extract elements of each column. Also, note that there is an extra newline character at the end of each print statement.

Reading CSV files

The program to read CSV files is similar to the previous program. A combination of strip method in the for-loop will be sufficient to produce the same result. But, as true Python programmers, we should use a function that is built specifically for this purpose.

```
>>> import csv
>>> for i in csv.reader(open('myfile.csv')):
...     print i
['Python', '2.7.1']
['Django', '1.4.1']
['Apache', '2.4']
```

Reading Excel files

Microsoft Excel files can be read using the `xlrd` module and written using the `xlwt` module. Here is a simple example of reading an Excel file using the `xlrd` module.

```
>>> import xlrd
>>> w = xlrd.open_workbook('myfile.xls')
>>> s = w.sheet_by_index(0)
>>> for rownumber in range(s.nrows):
...     print s.row_values(rownumber)
```

In line 2, the `open_workbook()` function is used to read the file. In line 3, the first sheet in the Excel file is obtained. In lines 4 and 5, the various lines in that sheet are iterated and its content printed.

User defined functions

A function allows reuse of code fragments. A Python function can be created using the def statement. Here is an example:

```
import math
def circleproperties(r):
    area = math.pi*r*r;
    circumference = 2*math.pi*r;
    return (area,circumference)


(a,c) = circleproperties(5) # Radius of the circle is 5
print "Area and Circumference of the circle are",a,c
```

The function circleproperties takes in one input argument, the radius (r). The return statement at the end of the function definition passes the computed values (in this case area and circumference) to the calling function, circleproperties. To invoke the function, use the name of the function and provide the radius value as argument enclosed in parenthesis. Finally, the area and circumference of the circle are displayed using the print command.

The variables area and circumference have local scope. Hence the variables cannot be invoked outside the body of the function. It is possible to pass on variables to a function that have global scope using the global statement.

1.6 Summary

- Python is a popular high level programming language. It is used for most common programming tasks such as scientific computation, text processing, dynamic website creation etc.

- Python distributions such as Enthought Python distribution or PythonXY are pre-built with many scientific modules and enables scientists to focus on their research instead of installation of modules.

- Python like other programming languages uses common relational and mathematical operators, comment statements, for-loops, if-else statements etc.

- To program like a Pythonista, use lists, sets, dictionary and tuples liberally.

- Python can read most of the common text formats like CSV, Microsoft Excel etc.

1.7 Exercises

1. If you are familiar with any other programming language, list the differences between that language and Python.

2. Write a Python program that will print numbers from 10 to 20 using for-loop.

3. Create a list of state names such as states = ['Minnesota','Texas','New York','Utah','Hawaii']. Add another entry 'California' to the end of the list. Then, print all the values of this list.

4. Print the content of list from Question 3 and also the corresponding index using the Python command `enumerate` in the `for-loop`.

5. Create a 2D list of size 3-by-3 with the following elements: $1, 2, 3 | 4, 5, 6 | 6, 7, 8$

6. It is easy to convert a list to set and viceversa. For example a list $'mylist = [1, 1, 2, 3, 4, 4, 5]'$ can be converted to set using the command `newset = set(mylist)`. The set can be converted back to list using `newlist = list(newset)`. Compare the contents of mylist and newlist. What do you infer?

7. Look up documentation for `join` method and join the content of the list ['Minneapolis','MN','USA'] and obtain the string 'Minneapolis, MN, USA.'

8. Consider the following Python code:

```
a = [1,2,3,4,2,3,5]
b = []
for i in a:
    if i>2:
        b.append(i)
print b
```

Rewrite the above code using list comprehension and reduce the number of lines.

Chapter 2

Computing using Python Modules

2.1 Introduction

We discussed the basics of Python in the previous chapter. We learned that Python comes with various built in batteries or modules. These batteries or modules perform various specialized operations. The modules can be used to perform computation, database management, web server etc. Since this book is focused on creating scientific applications, we limit our focus to Python modules that allow computation such as scipy, numpy, matplotlib, Python Imaging Library (PIL) and scikits. We discuss the relevance of each of these modules and explain their use with examples. We also discuss creation of new Python modules.

2.2 Python Modules

A number of scientific Python modules have been created and are available in the two Python distributions used in this book. Some of the most popular modules relevant to this book's scope are:

1. **numpy**: A powerful library for manipulating arrays and matrices.

2. **scipy**: Provides functions for performing higher order mathemat-

ical operations such as filtering, statistical analysis, image processing etc.

3. **matplotlib**: Provides functions for plotting and other forms of visualization.

4. **Python Imaging Library**: Provides functions for basic image reading, writing and processing.

5. **scikits**: An add-on package for scipy. The modules in scikits are meant to be added to scipy after development.

In the following sections, we will describe these modules in detail. Please refer to [5],[9],[40] to learn more.

2.2.1 Creating Modules

A module can be thought of as a Python file containing multiple functions with other optional components. All these functions share a common namespace, namely the name of the module file. For example, the following program is a legal Python module.

```
# filename: examplemodules.py
version = '1.0'

def printpi():
    print 'The value of pi is 3.1415'
```

A function named printpi and a variable called version was created in this module. The function performs the simple operation of printing the value of π.

2.2.2 Loading Modules

To load this module, use the following command in the Python command line or in a Python program. The "examplemodules" is the name of the module file.

```
>>> import examplemodules
```

Once the module is loaded, the function can be run using the command below. The second command prints the version number.

```
>>> examplemodules.printpi()
The value of pi is 3.1415
```

```
>>> examplemodules.version
'1.0'
```

The example module shown above has only one function. A module may contain multiple functions. In such case, it is not prudent to load all functions in the module using the simple "import modulename" as the process of loading all the functions is slow. There are other alternatives to this approach.

In the first example, all the functions in the datetime module are loaded even though we are interested only in obtaining the current date using date.today().

```
>>> import datetime
>>> datetime.date.today()
datetime.date(2013, 7, 4)
```

In the second example, only the neccesary function in the datetime module that is needed is loaded.

```
>>> from datetime import date
>>> date.today()
datetime.date(2013, 7, 4)
```

In the third example, we import all the functions in a given module using *. Once imported, the file name (in this case "date") that contains the function (in this case "today()") needs to be specified.

```
>>> from datetime import *
>>> date.today()
datetime.date(2012, 12, 9)
```

In the fourth example, we import a module (in this case numpy) and rename it to something shorter (such as np). This will reduce the amount of characters that need to be typed and consequently the lines of code to maintain.

```
>>> import numpy as np
>>> np.ones([3,3])
array([[ 1.,  1.,  1.],
       [ 1.,  1.,  1.],
       [ 1.,  1.,  1.]])
```

For the purpose of this book, we focus on only a few modules that are detailed below.

2.3 Numpy

A numpy module adds the ability to manipulate arrays and matrices using a library of high-level mathematical functions. Numpy is derived from the now defunct modules Numeric and Numarray. Numeric was the first attempt to provide the ability to manipulate arrays but it was very slow for computation on large arrays. Numarray, on the other hand, was too slow on small arrays. The code base was combined to create numpy.

Numpy has functions and routines to perform linear algebra, random sampling, polynomials, financial functions, set operations etc. Since this book is focused on image processing and since images are arrays, we will be using the matrix manipulation capabilities of numpy. The second module that we will be discussing is scipy, which internally uses numpy for its matrix manipulation.

The drawback of Python compared to C or C++ is the speed of execution. This is in part due to its interpreted execution. A Python program for numeric computation written with a similar construct to a C program using for-loop will perform considerably poorly. The best method of programming Python for faster execution is to use numpy and scipy modules. The following program illustrates the problem in programming using for-loop. In this program, the value of π is calculated using the Gregory-Leibiniz method. The method can be expressed as

$$\pi = 4 * \left\{ 1 - \frac{1}{3} + \frac{1}{5} - \frac{1}{7} + \frac{1}{9} \cdots \right\} \qquad (2.1)$$

The corresponding program is shown below. In the program, we perform the following operations:

1. Create the numerator and denominator separately using the linspace and ones functions. The details of the two functions can be found in the numpy documentation.

2. Begin a while-loop and find the ratio between the elements of numerator and denominator and the corresponding sum.

3. Multiply the value of the sum with 4 to obtain the value of π.

4. Print the time for completing the operation.

```
import time
from numpy import *

def main():
    noofterms = 10000000
    # Calculate the denominator.
    # First few terms are 1,3,5,7 ...
# den is short for denominator
    den = linspace(1,noofterms*2,noofterms)
```

```
# Calculate the numerator
# The first few terms are
# 1, -1, 1, -1 ...
# num is short for numerator
num = ones(noofterms)
for i in range(1,noofterms):
    num[i] = pow(-1,i)

counter = 0
sum_value = 0

t1 = time.clock()
while counter<noofterms:
    sum_value = sum_value+
              (num[counter]/den[counter])
    counter = counter + 1
pi_value = sum_value*4.0
print "pi_value = %f" % pi_value
t2 = time.clock()
# Determine the time for computation
timetaken = t2-t1
print "Timetaken = %f seconds" % timetaken


if __name__ == '__main__':
    main()
```

The program below is same as the one above except for step 3, where instead of calculating the sum of the ratio of the numerator and denominator using a while-loop or for-loop, we calculate using numpy's sum function.

```
from numpy import *
```

```
import time

def main():
    # No of terms in the series
    noofterms = 1000000
    # Calculate the denominator.
    # First few terms are 1,3,5,7 ...
    # den is short for denominator
    den  = linspace(1,noofterms*2,noofterms)
    # Calculate the numerator.
    # The first few terms are 1, -1, 1, -1 ...
    # num is short for numerator
    num = ones(noofterms)
    for i in range(1,noofterms):
        num[i] = pow(-1,i)
    print num
    print den
    # Find the ratio and sum all the fractions
    # to obtain pi value
    # Start the clock
    t1 = time.clock()
    pi_value =  sum(num/den)*4.0
    print "pi_value = %f" % pi_value
    t2 = time.clock()
    # Determine the time for computation
    timetaken = t2-t1
    print "Timetaken = %f seconds" % timetaken


if __name__ == '__main__':
    main()
```

The details of the sum function are given below:

```
numpy.sum(a, axis=None, dtype=None, out=None)


Necessary arguments :

a is the numpy array that needs to be summed.

Optional arguments:

axis can be None, integer or integer tuple.
This is the axis number over which the sum is calculated.
The default value is None and all elements will be summed.

dtype is the type of the array that is returned.
Default value is None.

Returns: An output ndarray of same shape as the input array
but with the specified axis removed. If a is a 2D array,
the output will be a 1D array.
```

The first program took 5.693210 seconds while the second program took 0.101271 seconds, for an approximate speed-up of 56. Although this example performs a fairly simple computation, a real world problem that takes a few weeks to solve can be completed in a few days with the appropriate use of numpy and scipy. Also, the program is elegant without the indentation used in the while loop.

2.3.1 Numpy Array or Matrices?

Numpy manipulates mathematical matrices and vectors and hence computes faster than a traditional for-loop that manipulate scalars. In numpy, there are two types of mathematical matrix classes: arrays and matrices. The two classes have been designed for similar purposes but arrays are more general-purpose n-dimensional while matrices facilitate faster linear algebra calculations. Some of the differences between arrays and matrices are listed below:

1. Matrix objects have rank 2, while arrays have rank > 2.

2. Matrix objects can be multiplied by using * operator while the same operator on an array performs element by element multiplication. The *dot()* needs to be used for performing multiplication on arrays.

3. Array is the default datatype on numpy.

The arrays are used more often in numpy and other modules that use numpy for their computation. The matrix and array can be interchanged but it is recommended to use arrays.

2.4 Scipy

Scipy is a library of programs and mathematical tools for scientific programming in Python. It uses numpy for its internal computation. Scipy is an extensive library that allows programming different mathematical applications such as Integration, Optimization, Fourier Transforms, Signal Processing, Statistics, Multi-dimensional image processing etc.

Travis Oliphant, Eric Jones and Pearu Peterson merged their modules to form scipy in 2001. Since then, many volunteers all over the world have participated in maintaining scipy.

As stated in Section 2.2, loading modules can be expensive both in CPU and memory used. This is especially true for large packages like scipy that contain many submodules. In such cases, load only the specific submodule.

```
>>> from scipy import ndimage
>>> import scipy.ndimage as im
```

In the first command, only the ndimage submodule is loaded. In the second command, the ndimage module is loaded as im.

This section is a brief introduction to scipy. The subsequent chapters will use scipy for all their image processing computations and hence details will be discussed later.

2.5 Matplotlib

Matplotlib is a 2D or a 3D plotting library for Python. It is designed to use numpy datatype. It can be used for generating plots inside a Python program. An example demonstrating the features of matplotlib is shown in Figure 2.5.

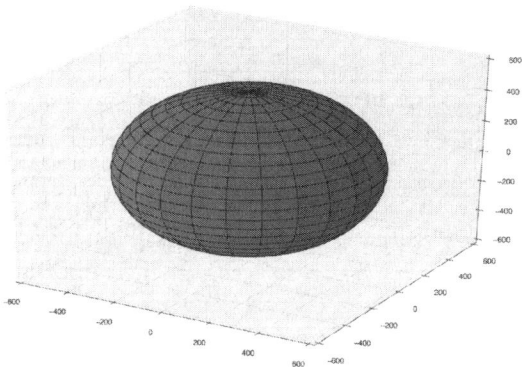

FIGURE 2.1: Example of a plot generated using matplotlib.

2.6 Python Imaging Library

Python Imaging Library (PIL) is a module for reading, writing and processing image files. It supports most of the common image formats like JPEG, PNG, TIFF etc. In a subsequent section, PIL will be used for reading and writing images.

2.7 Scikits

Scikits is a short form for scipy toolkits. It is an additional package that can be used along with scipy tools. An algorithm is programmed in scikits if

1. The algorithm is still under development and is not ready for prime time in scipy.

2. The package has a license that is not compatible with scipy.

3. Scipy is a general purpose scientific package in Python. Thus, it is designed so that it is applicable to a wide array of fields. If a package is deemed specialized for certain field, it continues to be part of scikits.

Scikits consists of modules from various fields such as environmental science, statistical analysis, image processing, microwave engineering, audio processing, boundary value problem, curve fitting etc.

In this book, we will focus only on the image processing routines in scikits named scikit-image. The scikit-image contains algorithms for input/output, morphology, object detection and analysis etc.

```
>>> from skimage import filter
>>> import skimage.filter as fi
```

In the first command, only the filter submodule is loaded. In the second command, the filter module is loaded as fi.

2.8 Python OpenCV Module

Open Source Computer Vision Library (OpenCV) is an image processing, computer vision and machine learning software library. It has more than 2000 algorithms for processing image data. It has a large user base and is used extensively in academic institutions, commercial organizations, and government agencies. It provides binding for common programming languages such as C, C++, Python etc. The Python binding is used in few examples in this book.

To import the Python OpenCV module, type the following in the command line:

```
>>> import cv2
```

2.9 Summary

- Various Python modules for performing image processing were discussed. They are numpy, scipy, matplotlib, Python Imaging Library, Python OpenCV, and scikits.

- The module has to be loaded before using functions that are specific to that module.

- In addition to using existing Python modules, user defined modules can be created.

- Numpy modules add the ability to manipulate arrays and matrices using a library of high-level mathematical functions. Numpy has two data structures for storing mathematical matrices. They are array and matrix. An array is more versatile than a matrix and is more commonly used in numpy and also in all the modules that use numpy for computation.

- Scipy is a library of programs and mathematical tools for scientific programming in Python.

- Scikits is used for the development of new algorithms that can later be incorporated into scipy.

2.10 Exercises

1. Python is an open-source and free software. Hence, there are many modules created for image processing. Perform research and discuss some of the benefits of each module over another.

2. Although this book is on image processing, it is important to combine the image processing operation with other mathematical operations such as optimization, statistics etc. Perform research about combining image processing with other mathematical operations.

3. Why is it more convenient to arrange the various functions as modules?

4. You are provided a CSV file containing a list of full path to file names of various images. The file has only one column with multiple rows. Each row contains the path to one file. You need to read the file name and then read the image as well. The method for reading a CSV file was shown in Chapter 1.

5. Modify the program from Question 4 to read a Microsoft Excel file instead.

6. Create a numpy array of size 5-by-5 containing all random values. Determine the transpose and inverse of this matrix.

Chapter 3

Image and its Properties

3.1 Introduction

We begin this chapter with an introduction to images, image types, and data structures in Python. Image processing operations can be imagined as a workflow similar to Figure 3.1. The workflow begins with reading an image. The output is then processed using either low-level or high-level operations. Low-level operations operate on individual pixels. Such operations include filtering, morphology, thresholding etc. High-level operations include image understanding, pattern recognition etc. Once processed, the images are either written to disk or visualized. The visualization may be performed during the course of processing as well. We will discuss this workflow and the functions using Python as an example.

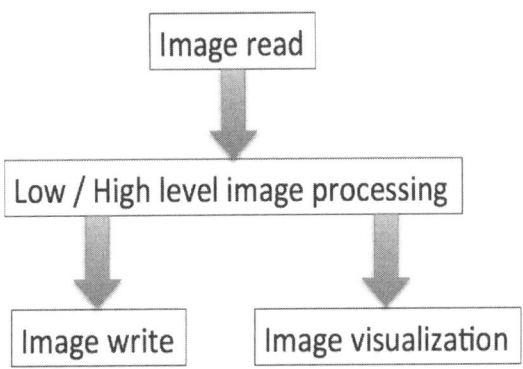

FIGURE 3.1: Image processing work flow.

3.2 Image and its Properties

In the field of medical imaging, the images may span all spatial domains and also the time domain. Hence it is common to find images in 3D and in some cases such as cardiac CT, images in 4D. In the case of optical microscopy, the images of the same specimen may be acquired at various emission and excitation wavelengths. Such images will span multiple channels and may have more than 4 dimensions. We begin the discussion by clarifying some of the mathematical terms that are used in this book.

For simplicity, let us assume the images that will be discussed in this book as 3D volumes. A 3D volume (I) can be represented mathematically as

$$\alpha = I \longrightarrow \mathbb{R} \text{ and } I \subset \mathbb{R}$$

Thus, every pixel in the image has a real number as its value. However, in reality as it is easier to store integers than to store floats, most images have integers for pixel values.

3.2.1 Bit Depth

The pixel range of a given image format is determined by its bit depth. The range is $[0, 2^{bitdepth-1}]$. For example, an 8-bit image will have a range of $[0, 2^8 - 1] = [0, 255]$. An image with higher bit depth needs more storage in disk and memory. Most of the common photographic formats such as jpeg, png etc. use 8-bit for storage and only have positive values.

Medical and microscope images use a higher bit depth, as scientific applications demand higher accuracy. A 16-bit medical image will have values in the range $[0, 65535]$ for a total number $65536 \ (= 2^{16})$ values. For images that have both positive and negative pixel values, the range is $[-32768, +32768]$. The total number of values in this case is 65536

($= 2^{16}$) or a bit-depth of 16. A good example of such an image is a CT DICOM image.

Scientific image formats store the pixel values at high precision not only for accuracy but also to ensure that physical phenomenon that it records is not lost. In CT, for example a pixel value of > 1000 indicates bone. If the image is stored in 8-bit, the pixel value of bone would be truncated at 255 and hence permanently lost. In fact, the most significant pixels in CT have intensity > 255 and hence needs larger bit depth.

There are a few image formats that store images at even higher bit-depth such as 32 or 64. For example, a jpeg image containing RGB (3 channels) will have a bit-depth of 8 for each channel and hence has a total bit-depth of 24. Similarly a tiff microscope image with 5 channels (say) with each channel at 16-bit depth will have a total bit-depth of 80.

3.2.2 Pixel and Voxel

A pixel in an image can be thought of as a bucket that collects light or electrons depending on the type of detector used. A single pixel in an image spans a distance in the physical world. For example in Figure 3.2, the arrows indicate the width and height of a pixel placed adjacent to three other pixels. In this case, the width and height of this pixel is 0.5 mm. Thus in a physical space, traversing a distance of 0.5 mm is equivalent to traversing 1 pixel in the pixel space. For all practical purpose, we can assume that detectors have square pixels i.e., the pixel width and pixel height are the same.

The pixel size could be different for different imaging modalities and different detectors. For example, the pixel size is greater for CT compared to micro-CT.

In medical and microscope imaging, it is more common to acquire 3D images. In such cases, the pixel size will have a third dimension,

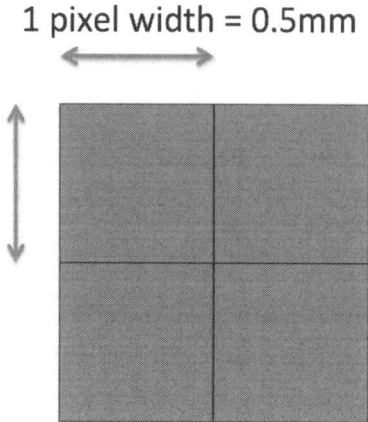

FIGURE 3.2: Width and height of pixel in physical space.

namely the pixel depth. The term pixel is generally applied to 2D and is replaced by voxel in 3D images.

Most of the common image formats like DICOM, nifti and some microscope image formats contain the voxel size in their header. Hence, when such images are read in a visualization or image processing program, an accurate analysis and visualization can be performed. But if the image does not have the information in the header or if the visualization or image processing program cannot read the header properly, it is important to use the correct voxel size for analysis.

Figure 3.3 illustrates the problem of using incorrect voxel size in visualization. The left image is the volume rendering of an optical coherence tomography image with incorrect voxel size. The right image is the volume rendering of the same image with correct voxel size. In the left image, it can be seen clearly that the object is highly elongated in the z-direction. In addition, the undulations at the top of the volume and the five hilly structures at the top are also made prominent by the incorrect voxel size. The right image has the same shape and size as the original object. The problem not only affects visualization but also any measurements performed on the volume.

 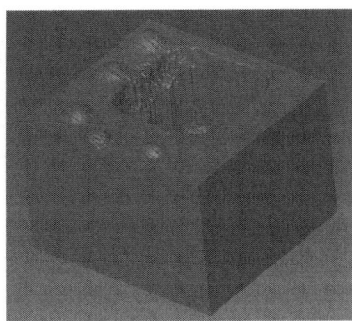

(a) Volume rendering with incorrect voxel size. The 3D is elongated in the z direction.

(b) Volume rendering with correct voxel size.

FIGURE 3.3: An example of volume rendering.

3.2.3 Image Histogram

A histogram is a graphical depiction of the distribution of pixel value in an image. The image in Figure 3.4 is a histogram of an image. The x-axis is the pixel value and the y-axis is the frequency or the number of pixels with the given pixel value. In the case of a integer based image such as jpeg, whose values spans $[0, 255]$, the number of values in the x-axis will be 256. Each of these 256 values is referred to as "bins." A few numbers of bins can also be used in the x-axis. In the case of images containing floating-point values, the bins will have a range of values.

Histograms are a useful tool in determining the quality of the image. A few observations can be made by using Figure 3.4:

1. The left side of the histogram corresponds to lower pixel values. Hence if the frequency at lower pixel values is very high, it indicates that some of the pixels might be missing from that end i.e., there are values to the further left of the first pixel that were not recorded in the image.

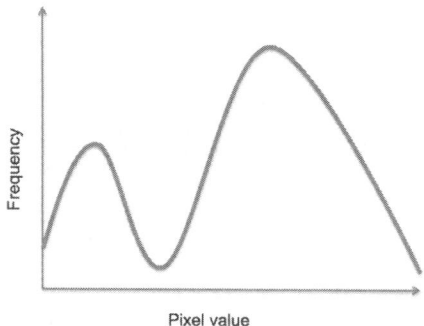

FIGURE 3.4: An example of a histogram.

2. The right side of the histogram corresponds to higher pixel values. Hence if the frequency at higher pixel values is very high, it indicates saturation.

3. The above histogram is bi-modal. The trough between the two peaks is the pixel value that can be used for segmentation by thresholding. But not all images have bi-modal histograms; hence there are many techniques for segmentation using histograms. We will discuss some of these techniques in the Chapter 7, Segmentation.

3.2.4 Window and Level

The human eye can view a large range of intensity values, while modern displays are severely limited in their capabilities.

Image viewing applications display the pixel value after a suitable transformation due to the fact that displays have a lower intensity range than the intensity range in an image. One example of the transformation, namely window-level, is shown in Figure 3.5. Although the computer selects a transformation, the user can modify it by changing the window range and the level. The window allows modifying the contrast of the display while the level changes the brightness of the display.

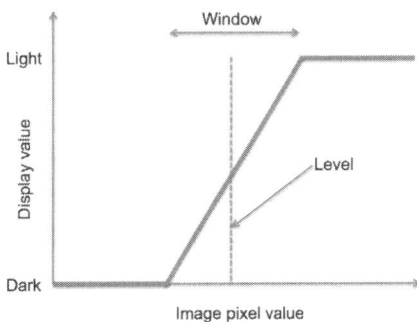

FIGURE 3.5: Window and level.

3.2.5 Connectivity: 4 or 8 Pixels

The usefulness of this section will be more apparent with the discussion of convolution in Chapter 6, Fourier Transform. During the convolution operation, a mask or kernel is placed on top of an image pixel. The final value of the output image pixel is determined using a linear combination of the value in the mask and the pixel value in the image. The linear combination can be calculated for either 4-connected pixels or 8-connected pixels. In the case of 4-connected pixels shown in Figure 3.6, the process is performed on the top, bottom, left and right pixels. In the case of 8-connected pixels, the process is performed in addition on the top-left, top-right, bottom-left and bottom-right pixels.

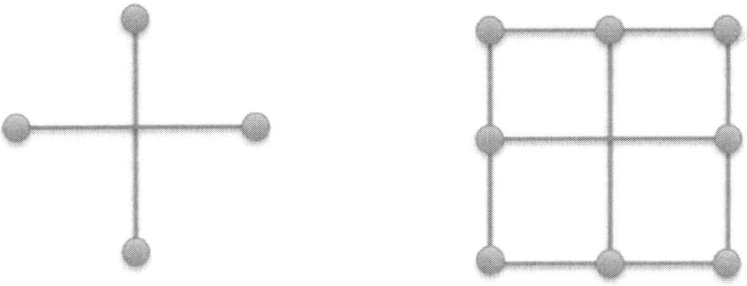

FIGURE 3.6: An example of 4 and 8 pixel connectivity.

3.3 Image Types

There are more than 100 image formats. Some of these formats such as jpeg, gif, png etc. are used for photographic images. Formats such as DICOM, nifti, analyze avw are used in medical imaging. Formats such as tiff, ics, ims etc. are used in microscope imaging. In the following sections, we discuss some of these formats.

3.3.1 JPEG

JPEG stands for the Joint Photographic Experts Group, a joint committee formed to add images to text terminals. Its extension is .jpg or .jpeg. It is one of the most popular formats due to its ability to compress the data significantly with minimal visual loss. In the initial days of the World Wide Web, jpeg became popular as it helped save bandwidth in image data transfer. It is a lossy format, that compresses data using Discrete Cosine Transform (DCT). The parameters of compression can be tuned to minimize the loss in details. Since jpeg stores image data after transforming them using DCT, it is not very suitable for storing images that contain fine structures such as lines, curves etc. Such images are better stored as png or tif. The jpeg images can be viewed using viewers built in to most computers. Since jpeg images can be compressed, image standards such as tiff and DICOM may use jpeg compression to store the image data when compression is needed.

3.3.2 TIFF

TIFF stands for Tagged Image File Format. Its extension is .tif or .tiff. The latest version of the tif standards is 6.0. It was created in the 80's for storing and encoding scanned documents. It was developed by Aldus Corporation, which was later acquired by Adobe Systems. Hence, the copyright for tiff standards is held by Adobe Systems.

Originally it was developed for single bit data but today's standards allow storage of 16 bit and even floating point data. Charged Couple Device (CCD) cameras used in scientific experiments acquire images at more than 12 bit resolution and hence tif images that store high precision are used extensively. The tif images can be stored internally using jpeg lossy compression or can be stored with loseless compression such as LZW.

It is popular in the microscope community for the fact that it has higher bit depth (> 12 bits) per pixel per channel and also for its ability to store a sequence of images in a single tif file. The latter is sometimes referred as 3D tif. Most of the popular image processing software for the microscope community can read most forms of tif images. Simple tif images can be viewed using viewers built in to most computers. The tif images generated from scientific experiments are best viewed using applications that are specialized in that domain.

3.3.3 DICOM

Digital Imaging and Communication in Medicine (DICOM) is a standard format for encoding and transmitting medical CT and MRI data. This format stores the image information along with other data like patient details, acquisition parameters etc. DICOM images are used by doctors in various disciplines such as Radiology, Neurology, Surgery, Cardiology, Oncology etc. There are more than 20 DICOM committees that meet and update the standards 4 or 5 times a year. It is managed by the National Electrical Manufacturers Association (NEMA), which owns the copyright of the DICOM standards.

DICOM format uses tested tools such as JPEG, MPEG, TCP/IP for its internal working. This allows easier deployment and creation of DICOM tools. DICOM standards also define the transfer of images, storage and other allied workflow. Since DICOM standards have become popular, many image processing readers and viewers have been created to read, process and write images.

DICOM images have header as well as image data similar to other image formats. But, unlike other format header's, the DICOM header contains not only information about the size of the image, pixel size etc. but also patient information, physician information, imaging technique parameters etc. The image data may be compressed using various techniques like jpeg, lossless jpeg, run length encoding (RLE) etc. Unlike other formats, DICOM standards define both the data format and also the protocol for transfer.

The listing below is a partial example of a DICOM header. The patient and doctor information have been either removed or altered for privacy. Section 0010 contains patient information, section 0009 details the CT machine used for acquiring the image, and section 0018 details the parameter of acquisition etc.

```
0008,0022    Acquisition Date: 20120325
0008,0023    Image Date: 20120325
0008,0030    Study Time: 130046
0008,0031    Series Time: 130046
0008,0032    Acquisition Time: 130105
0008,0033    Image Time: 130108
0008,0050    Accession Number:
0008,0060    Modality: CT
0008,0070    Manufacturer: GE MEDICAL SYSTEMS
0008,0080    Institution Name: --------------------
0008,0090    Referring Physician's Name: XXXXXXX
0008,1010    Station Name: CT01_OCO
0008,1030    Study Description: TEMP BONE/ST NECK W
0008,103E    Series Description: SCOUTS
0008,1060    Name of Physician(s) Reading Study:
0008,1070    Operator's Name: ABCDEF
0008,1090    Manufacturer's Model Name: LightSpeed16
0009,0010    ---: GEMS_IDEN_01
0009,1001    ---: CT_LIGHTSPEED
0009,1002    ---: CT01
```

0009,1004	---: LightSpeed16
0010,0010	Patient's Name: XYXYXYXYXYXYX
0010,0020	Patient ID: 213831
0010,0030	Patient's Birth Date: 19650224
0010,0040	Patient's Sex: F
0010,1010	Patient's Age:
0010,21B0	Additional Patient History: ? MASS RIGHT EUSTACHIAN TUBE
0018,0022	Scan Options: SCOUT MODE
0018,0050	Slice Thickness: 270.181824
0018,0060	kVp: 120
0018,0090	Data Collection Diameter: 500.000000
0018,1020	Software Versions(s): LightSpeedverrel
0018,1030	Protocol Name: 3.2 SOFT TISSUE NECK
0018,1100	Reconstruction Diameter:
0018,1110	Distance Source to Detector: 949.075012
0018,1111	Distance Source to Patient: 541.000000
0018,1120	Gantry/Detector Tilt: 0.000000
0018,1130	Table Height: 157.153000
0018,1140	Rotation Direction: CW
0018,1150	Exposure Time: 2772
0018,1151	X-ray Tube Current: 10
0018,1152	Exposure: 27
0018,1160	Filter Type: BODY FILTER
0018,1170	Generator Power: 1200
0018,1190	Focal Spot(s): 0.700000
0018,1210	Convolution Kernel: STANDARD

The various software that can be used to manipulate DICOM images can be found online. We will classify these software based on the user requirements.

The user might need:

1. A simple viewer with limited manipulation like ezDICOM.

2. A viewer with ability to manipulate images and perform rendering like Osirix.

3. A viewer with image manipulation capability and also extensible with plugins like ImageJ.

ezDICOM: This is a viewer that provides sufficient functionality that allows users to view and save DICOM files without installing any other complex software in their system. It is available only for Windows OS. It can read DICOM files and save them in other file formats. It can also convert image files to Analyze format. It is a available at http://www.mccauslandcenter.sc.edu/mricro/ezdicom/.

Osirix: This is a viewer with extensive functionality and is available free, but unfortunately it is available only in MacOSX. Like other DICOM viewers, it can read and store files in different file formats and as movies. It can perform Multi-Planar Reconstruction (MPR), 3D surface rendering, 3D volume rendering, and endoscopy. It can also view 4D DICOM data. The surface rendered data can also be stored as VRML, STL files etc. It is a available at http://www.osirix-viewer.com/.

ImageJ: ImageJ was funded by the National Institutes of Health (NIH) and is available as open source. It is written in Java and users can add their own Java classes or plugins. It is available in all major operating system like Windows, Linux, UNIX, Mac etc. It can read all DICOM formats and can store the data in various common file formats and also as movies. The plugins allow various image processing operations. Since the plugins can be easily added, the complexity of the image processing operation is limited only by the user's knowledge of Java. Since ImageJ is a popular image processing software, a brief introduction is presented in Appendix C, Introduction to ImageJ.

3.4 Data Structures for Image Analysis

Image data is generally stored as a mathematical matrix. So in general, a 2D image of size 1024-by-1024 is stored in a matrix of the same size. Similarly, a 3D image is stored in a 3D matrix. In numpy, a mathematical matrix is called a numpy array. As we will be discussing in the subsequent chapters, the images are read and stored as numpy array and then processed using either functions in a Python module or user-defined functions.

Since Python is a dynamically typed language (i.e., no defining data type), it will determine the data type and size of the image at run time and store appropriately.

3.4.1 Reading Images

There are a few different methods for reading common image formats like jpeg, tif, png etc. We will discuss a couple of them. In the subsequent chapters, we will continue to use the same approach for reading images.

```
import scipy.misc as mi
im = mi.imread('Picture1.png')

from scipy.misc.pilutil import Image
im = Image.open('Picture1.png')
```

Reading DICOM images using pyDICOM

We will use pyDICOM (http://code.google.com/p/pyDICOM/) a module in Python to read or write or manipulate DICOM images. The process for reading DICOM images is similar to jpeg, png etc. Instead of using scipy, the pyDICOM module is used. The pyDICOM module is not installed by default in the two distributions. The instruction for installing the modules in the two distributions on Windows, Mac or

Linux is provided in the Appendix A, Installing Python distributions. To read a DICOM file, the DICOM module is first imported. The file is then read using "read_file" function.

```
import dicom
ds = dicom.read_file("ct_abdomen.dcm")
```

3.4.2 Displaying Images

To display images in Mac, an environment variable has to be set using the instruction below. The image can then be viewed using the imshow command in the program.

```
export SCIPY_PIL_IMAGE_VIEWER =
 /Applications/Preview.app/Contents/MacOS/Preview


To check type 'env | grep -i scipy' in the
Linux / Mac command line


import scipy.misc as mi
im = mi.imread('Picture1.png')
mi.imshow(im)
```

3.4.3 Writing Images

There are few different methods for writing or saving common image formats like jpeg, tif, png etc. We will discuss couple of them. In the first method, the scipy.misc module is used.

```
# importing scipy.misc as mi
import scipy.misc as mi

# reading the image
im = mi.imread('Picture1.png')
# saving the image
mi.imsave('test.png',im)
```

In the second method, the scipy.misc.pilutil module is used.

```
# importing Image module
from scipy.misc.pilutil import Image

# opening and reading the image
im = Image.open('Picture1.png')
# saving the image
im.save('test.png')
```

In the subsequent chapters, we will continue to use these approaches for writing or saving images.

Writing DICOM images using pyDICOM

To write a DICOM file, the DICOM module is first imported. The file is then written using "write_file" function. The input to the function is the name of the DICOM file and also the array that needs to be stored.

```
import dicom
datatowrite = ...
dicom.write_file("ct_abdomen.dcm",datatowrite)
```

3.5 Programming Paradigm

As described in the introductory section, the workflow (Figure 3.1) for image processing begins with reading an image and finally ends with either writing the image to file or visualizing it. The image processing operations are performed between the reading and writing or visualizing of the image. In this section, the code snippet that will be used for reading and writing or visualizing of the image is presented. This code snippet will be used in all the programs presented in this book.

```
# scipy.misc module along with fromimage
```

```
# and toimage is used to convert
# an image to numpy array and vice versa
import scipy.misc
# Image is imported to open and read the image
from scipy.misc.pilutil import Image

# opening and reading the image
# converting the image to grayscale
im = Image.open('Picture1.png').convert('L')
# im is converted to a numpy ndarray
im = scipy.misc.fromimage(im)
# Peform image processing computation/s on im and
# store the results in b
# b is converted to an image
c = scipy.misc.toimage(b)
# c is saved as test.png
c.save('test.png')
```

In the first line, the scipy.misc module is imported. In the second line, Image function is imported from scipy.misc.pilutil. In the fourth line, the image is opened using Image.open function. If the image is RGB and needs to be converted to grayscale, the function convert('L') needs to be used for grayscale conversion. However, if the image is already a grayscale image then the convert function is not required. The line "Perform image processing computation/s" will be replaced with appropriate image processing operation/s. Once the image processing operation is complete, the result is stored in the variable b (say). This variable is a numpy n-dimensional array, ndarray and needs to be converted to an image before it can be saved or visualized. This is accomplished by using the scipy.misc.toimage() function. Finally the image is saved using the function save(). Python provides multiple functions through its numerous modules for performing this operation. For simplicity, this method was chosen.

3.6 Summary

- Image processing is preceded by reading an image file. It is then followed by either writing the image to file or visualization.

- Image is stored generally in the form of matrices. In Python, it is processed as a numpy n-dimensional array, ndarray.

- Image has various property like bit-depth, pixel/voxel size, histogram, window-level etc. These properties affect the visualization and processing of images.

- There are hundreds of image formats created to serve the needs of image processing community. Some of these formats like jpeg, png etc. are used generally for photographs while DICOM, avw, nifti are used for medical image processing.

- In addition to processing these images, it is important to view these images using graphical tools such as ezDicom, Osirix, ImageJ etc.

- Reading and writing images can be performed using many methods. One such method was presented in this chapter. We will continue to use this method in all the subsequent chapters.

3.7 Exercises

1. An image of size 100-by-100 has isotropic pixel size of 2-by-2 microns. The number of pixels in the foreground is 1000. What is the area of the foreground and background in $microns^2$?

2. A series of images are used to create a volume of data. There are 100 images each of size 100-by-100. The voxel size is 2-by-2-by-2

microns. Determine the volume of the foreground in $microns^3$ given the number of pixels in the foreground is 10,000.

3. A histogram plots the frequency of occurrence of the various pixel values. This plot can be converted to a probability density function or pdf, so that the y-axis is the probability of the various pixel values. How can this be accomplished?

4. To visualize window or level, open an image in any image processing software (such as ImageJ). Adjust window and level. Comment on the details that can be seen for different values of window and level.

5. There are specialized formats for microscope images. Conduct research on these formats.

Part II

Image Processing using Python

Chapter 4

Spatial Filters

4.1 Introduction

So far we have covered the basics of Python and its scientific modules. In this chapter, we begin our journey of learning image processing. The first concept we will master is filtering, which is at the heart of image enhancement. Filters are used to remove noise or undesirable impurities. The first derivative and second derivative filters are used to determine edges in an image.

There are two types of filters: linear filters and non-linear filters. Linear filters include mean, Laplacian and Laplacian of Gaussian. Non-linear filters include median, maximum, minimum, Sobel, Prewitt and Canny filters.

Image enhancement can be accomplished in two domains: spatial and frequency. The spatial domain constitutes all the pixels in an image. Distances in the image (in pixels) correspond to real distances in micrometers, inches etc. The domain over which the Fourier transformation of an image ranges is known as the frequency domain of the image. We begin with image enhancement techniques on spatial domain. Later in Chapter 6, Fourier Transform, we will discuss image enhancement using frequency or Fourier domain.

The Python modules that are used in this chapter are scikits and scipy. Scipy documentation can be found at [90], scikits documentation can be found at [88], and scipy ndimage documentation can be found at [91].

4.2 Filtering

Filtering is a commonly used tool in image processing. As a water filter removes impurities, an image processing filter removes undesired features (such as noise) from an image. Each filter has a specific utility and is designed to either remove a type of noise or to enhance certain aspects of the image. We will discuss many filters along with their purposes and their effects on images.

For filtering, a filter or mask is used. It is usually a two dimensional square window that moves across the image affecting only one pixel at a time. Each number in the filter is known as a coefficient. The coefficients in the filter determine the effects of the filter and consequently the output image. Let us consider a 3-by-3 filter, F, given in Table 4.1.

F_1	F_2	F_3
F_4	F_5	F_6
F_7	F_8	F_9

TABLE 4.1: A 3-by-3 filter.

If (i, j) is the pixel in the image, then a sub-image around (i, j) of the same dimension as the filter is considered for filtering. The center of the filter should overlap with (i, j). The pixels in the sub-image are multiplied with the corresponding coefficients in the filter. This yields a matrix of the same size as the filter. The matrix is simplified using a mathematical equation to obtain a single value that will replace the pixel value in (i, j) of the image. The exact nature of the mathematical equation depends on the type of filter. For example, in the case of a mean filter, the value of $F_i = \frac{1}{N}$, where N is the number of elements in the filter. The filtered image is obtained by repeating the process of placing the filter on every pixel in the image, obtaining the single value and replacing the pixel value in the original image. This process of sliding a filter window over an image is called convolution in the spatial domain.

Let us consider the following sub-image from the image, I, centered at (i, j)

$I(i-1, j-1)$	$I(i-1, j)$	$I(i-1, j+1)$
$I(i, j-1)$	$I(i, j)$	$I(i, j+1)$
$I(i+1, j-1)$	$I(i+1, j)$	$I(i+1, j+1)$

TABLE 4.2: A 3-by-3 sub-image.

The convolution of the filter given in Table 4.1 with the sub-image in Table 4.2 is given as follows:

$$\begin{aligned}
I_{new}(i,j) = F_1 * I(i-1, j-1) + F_2 * I(i-1, j) + F_3 * I(i-1, j+1) \\
+ F_4 * I(i, j-1) + F_5 * I(i, j) + F_6 * I(i, j+1) \\
+ F_7 * I(i+1, j-1) + F_8 * I(i+1, j) + F_9 * I(i+1, j+1)
\end{aligned} \quad (4.1)$$

where $I_{new}(i, j)$ is the output value at location (i, j). This process has to be repeated for every pixel in the image. Since the filter plays an important role in the convolution process, the filter is also known as the convolution kernel.

The convolution operation has to be performed at every pixel in the image including pixels at the boundary of the image. When the filter is placed on the boundary pixels, a portion of the filter will lie outside the boundary. Since the pixel values do not exist outside the boundary, new values have to be created prior to convolution. This process of creating pixel values outside the boundary is called padding. The padded pixels can be assumed to be either zero or a constant value. Other padding options such as nearest neighbor or reflect create padded pixels using pixel values in the image. In the case of zeros, the padded pixels are all zeros. In the case of constant, the padded pixels take a specific value. In the case of reflect, the padded pixels take the value of the last row/s or column/s. The padded pixels are considered only for convolution and will be discarded after convolution.

Let us consider an example to show different padding options. Figure 4.1(a) is a 7-by-7 input image that will be convolved using a 3-by-5 filter with the center of the filter at $(1, 2)$. In order to include boundary pixels for convolution, we pad the image with one row above and one row below and two columns to the left and two columns to the right. In general the size of the filter dictates the number of rows and columns that will be padded to the image.

- **Zero padding**: All padded pixels are assigned a value of zero (Figure 4.1(b)).

- **Constant padding**: A constant value of 5 is used for all padded pixels (Figure 4.1(c)). The constant value can be chosen based on the type of image that is being processed.

- **Nearest neighbor**: The values from the last row or column (Figure 4.1(d)) are used for padding.

- **Reflect**: The values from the last row or column (Figure 4.1(e)) are reflected across the boundary of the image.

4.2.1 Mean Filter

In mathematics, functions are classified into two groups, linear and non-linear. A function f is said to be linear if

$$f(x+y) = f(x) + f(y) \qquad (4.2)$$

Otherwise, f is non-linear. A linear filter is an extension of the linear function.

An excellent example of a linear filter is the mean filter. The coefficients of mean filter F (Table 4.1) are 1's. To avoid scaling the pixel intensity after filtering, the whole image is then divided by the number

0	2	5	7	3	10	9
11	1	4	6	8	2	0
0	12	10	9	7	4	5
1	9	7	8	13	11	0
5	10	14	6	2	1	1
7	6	11	3	13	8	4
3	9	6	12	7	10	5

(a) A 7-by-7 input image.

0	0	0	0	0	0	0	0	0	0	0
0	0	0	2	5	7	3	10	9	0	0
0	0	11	1	4	6	8	2	0	0	0
0	0	0	12	10	9	7	4	5	0	0
0	0	1	9	7	8	13	11	0	0	0
0	0	5	10	14	6	2	1	1	0	0
0	0	7	6	11	3	13	8	4	0	0
0	0	3	9	6	12	7	10	5	0	0
0	0	0	0	0	0	0	0	0	0	0

(b) Padding with zeros.

5	5	5	5	5	5	5	5	5	5	5
5	5	0	2	5	7	3	10	9	5	5
5	5	11	1	4	6	8	2	0	5	5
5	5	0	12	10	9	7	4	5	5	5
5	5	1	9	7	8	13	11	0	5	5
5	5	5	10	14	6	2	1	1	5	5
5	5	7	6	11	3	13	8	4	5	5
5	5	3	9	6	12	7	10	5	5	5
5	5	5	5	5	5	5	5	5	5	5

(c) Padding with a constant.

0	0	0	2	5	7	3	10	9	9	10
0	0	0	2	5	7	3	10	9	9	9
11	11	11	1	4	6	8	2	0	0	0
0	0	0	12	10	9	7	4	5	5	5
1	1	1	9	7	8	13	11	0	0	0
5	5	5	10	14	6	2	1	1	1	1
7	7	7	6	11	3	13	8	4	4	4
3	3	3	9	6	12	7	10	5	5	5
3	3	3	9	6	12	7	10	5	5	5

(d) Padding with nearest neighbor.

2	0	0	2	5	7	3	10	9	9	10
2	0	0	2	5	7	3	10	9	9	10
1	11	11	1	4	6	8	2	0	0	2
12	0	0	12	10	9	7	4	5	5	4
9	1	1	9	7	8	13	11	0	0	11
10	5	5	10	14	6	2	1	1	1	1
6	7	7	6	11	3	13	8	4	4	8
9	3	3	9	6	12	7	10	5	5	10
9	3	3	9	6	12	7	10	5	5	10

(e) Padding with reflect option.

FIGURE 4.1: An example of different padding options.

of pixels in the filter; in the case of a 3-by-3 subimage we divide it by 9.

Unlike other filters discussed in this chapter, the mean filter does not have a scipy.ndimage module function. However, we can use the convolve function to achieve the intended result. The following is the Python function for convolve:

```
scipy.ndimage.filters.convolve(input, weights)

Necessary arguments:
  input is the input image as an ndarray.

weights is an ndarray consisting of
coefficients of 1s for the mean filter.

Optional arguments:
  mode determines the method for handling the array
border by padding. Different options are: constant,
reflect, nearest, mirror, wrap.

  cval is a scalar value specified when the mode option
is constant. The default value is 0.0.

  origin is a scalar that determines filter origin.
The default value 0 corresponds to a filter
whose origin (reference pixel) is at the center.
In a 2D case, origin = 0 would mean (0,0).

Returns: output is an ndarray
  For example, a 100-by-100 image is 2 dimensional.
```

The program explaining the usage of the mean filter is given below.

Spatial Filters

The filter (k) is a numpy array of size 5-by-5 with all values = 1/25. The filter is then used for convolution using the "convolve" function from scipy.ndimage.filters.

```
import numpy as np
import scipy.ndimage
from scipy.misc.pilutil import Image

# opening the image and converting it to grayscale
a = Image.open('../Figures/ultrasound_muscle.png').
    convert('L')
# initializing the filter of size 5 by 5
# the filter is divided by 25 for normalization
k = np.ones((5,5))/25
# performing convolution
b = scipy.ndimage.filters.convolve(a, k)
# b is converted from an ndarray to an image
b = scipy.misc.toimage(b)
b.save('../Figures/mean_output.png')
```

Figure 4.2(a) is an ultrasound image of muscle. Notice that the image contains noise. The mean filter of size 5-by-5 is applied to remove the noise. The output is shown in Figure 4.2(b). The mean filter effectively removed the noise but in the process blurred the image.

Advantages of the mean filter

- Removes noise.

- Enhances the overall quality of the image, i.e. mean filter brightens an image.

Disadvantages of the mean filter

- In the process of smoothing, the edges get blurred.

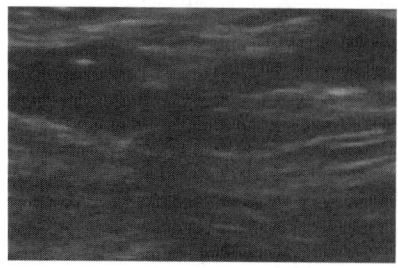 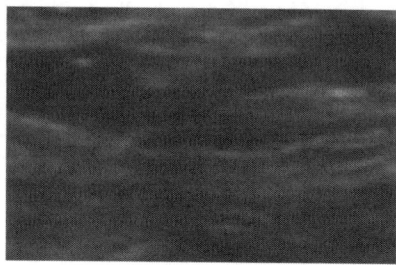

(a) Input image for mean filter. (b) Output generated with a filter size (5,5).

FIGURE 4.2: Example of mean filter.

- Reduces the spatial resolution of the image.

If the coefficients of the mean filter are not all 1s, then the filter is a weighted mean filter. In the weighted mean filter, the filter coefficients are multiplied with the sub-image as in the non-weighted filter. After application of the filter, the image should be divided by the total weight for normalization.

4.2.2 Median Filter

Functions that do not satisfy Equation 4.2 are non-linear. Median filter is one of the most popular non-linear filters. A sliding window is chosen and is placed on the image at the pixel position (i,j). All pixel values under the filter are collected. The median of these values is computed and is assigned to (i,j) in the filtered image. For example, consider a 3-by-3 sub-image with values 5, 7, 6, 10, 13, 15, 14, 19, 23. To compute the median, the values are arranged in ascending order, so the new list is: 5, 6, 7, 10, 13, 14, 15, 19, and 23. Median is a value that divides the list into two equal halves; in this case it is 13. So the pixel (i,j) will be assigned 13 in the filtered image. The median filter is most commonly used in removing salt-and-pepper noise and impulse noise. Salt-and-pepper noise is characterized by black and white spots randomly distributed in an image.

The following is the Python function for Median filter:

```
scipy.ndimage.filters.median_filter(input, size=None,
   footprint=None, mode='reflect', cval=0.0, origin=0)

Necessary arguments:
  input is the input image as an ndarray.

Optional arguments:
  size can be a scalar or a tuple. For example, if the
  image is 2D, size = 5 implies a 5-by-5 filter is
  considered. Alternately, size=(5,5) can also be specified.

  footprint is a boolean array of the same dimension as
  the size unless specified otherwise. The pixels in the
  input image corresponding to the points to the
  footprint with true values are considered for
  filtering.

  mode determines the method for handling the array
  border by padding. Options are: constant,
  reflect, nearest, mirror, wrap.

  origin is a scalar that determines origin of the
  filter. The default value 0 corresponds to a filter
  whose origin (reference pixel) is at the center. In a
  2D case, origin = 0 would mean (0,0).

Returns: output is an ndarray.
```

The Python code for median filter is given below:

```
import scipy.misc
import scipy.ndimage
from scipy.misc.pilutil import Image

# opening the image and converting it to grayscale
a = Image.open('../Figures/ct_saltandpepper.png').
    convert('L')
# performing the median filter
b = scipy.ndimage.filters.median_filter(a,size=5,
    footprint=None,output=None,mode='reflect',
    cval=0.0,origin=0)
# b is converted from an ndarray to an image
b = scipy.misc.toimage(b)
b.save('../Figures/median_output.png')
```

In the above code, $size = 5$ represents a filter (mask) of size 5-by-5. The image in Figure 4.3(a) is a CT slice of the abdomen with salt-and-pepper noise. Since the output image is an n dimensional ndarray, the scipy.misc.toimage command is used to convert the array into an image for visualization and saving purposes. The output image is shown in Figure 4.3(b). The median filter efficiently removed the salt-and-pepper noise.

4.2.3 Max Filter

This filter enhances the bright points in an image. In this filter the maximum value in the sub-image replaces the value at (i,j). The Python function for the maximum filter has the same arguments as the median filter discussed above. The Python code for the max filter is given below.

```
import scipy.misc
```

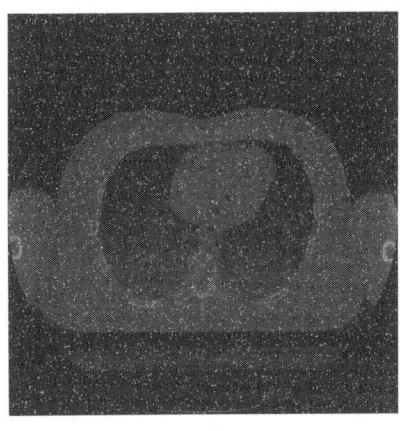 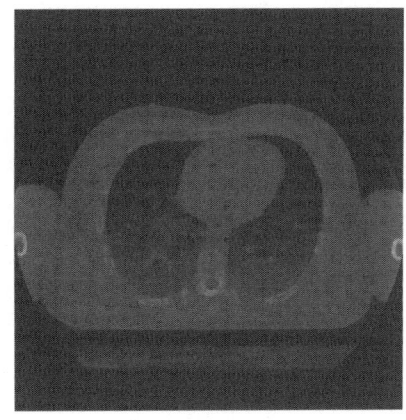

(a) Input image for median filter. (b) Output generated with a filter size=(5,5).

FIGURE 4.3: Example of median filter.

```
import scipy.ndimage
from scipy.misc.pilutil import Image

# opening the image and converting it to grayscale
a = Image.open('../Figures/wave.png').convert('L')
# performing maximum filter
b = scipy.ndimage.filters.maximum_filter(a,size=5,
    footprint=None,output=None,mode='reflect',
    cval=0.0,origin=0)
# b is converted from an ndarray to an image
b = scipy.misc.toimage(b)
b.save('../Figures/maxo.png')
```

The image in Figure 4.4(a) is the input image for the max filter. The input image has a thin black boundary at the left, right and bottom. After application of the max filter, the white pixels have grown and hence the thin edges in the input image are replaced by white pixels in the output image as shown in Figure 4.4(b).

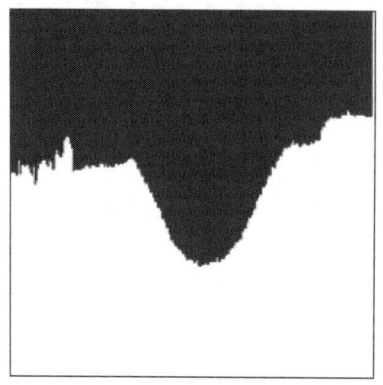 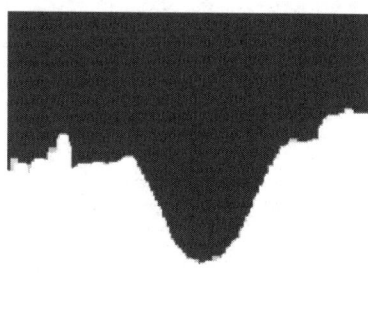

(a) Input image for max filter.   (b) Output image of max filter.

FIGURE 4.4: Example of max filter.

4.2.4   Min Filter

This filter is used to enhance the darkest points in an image. In this filter, the minimum value of the sub-image replaces the value at (i, j). The Python function for the minimum filter has the same arguments as the median filter discussed above. The Python code for the min filter is given below.

```
import scipy.misc
import scipy.ndimage
from scipy.misc.pilutil import Image

# opening the image and converting it to grayscale
a = Image.open('../Figures/wave.png').convert('L')
# performing minimum filter
b = scipy.ndimage.filters.minimum_filter(a,size=5,
    footprint=None,output=None,mode='reflect',
```

```
    cval=0.0,origin=0)
# b is converted from an ndarray to an image
b = scipy.misc.toimage(b)
# saving b as mino.png
b.save('../Figures/mino.png')
```

After application of the min filter to the input image in Figure 4.5(a), the black pixels have grown and hence the thin edges in the input image are thicker in the output image as shown in Figure 4.5(b).

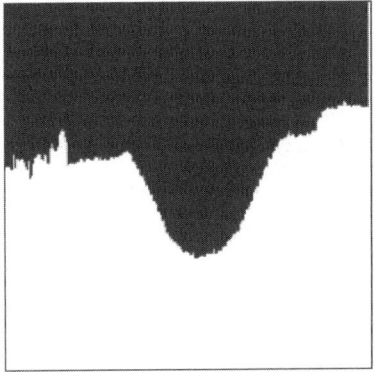
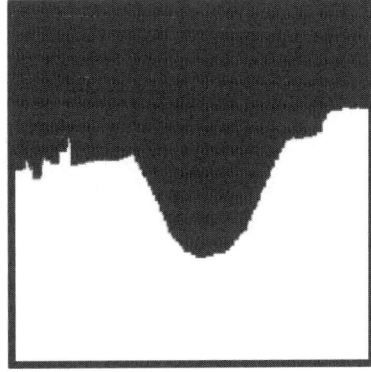

(a) Input image for min filter.   (b) Output image of min filter.

FIGURE 4.5: Example of min filter.

4.3 Edge Detection using Derivatives

Edges are a set of points in an image where there is a change of intensity. From calculus, we know that the changes in intensity can be measured by using the first or second derivative. First, let us learn how changes in intensities affect the first and second derivatives by considering a simple image and its corresponding profile. This method will form the basis for using first and second derivative filters for edge detection. Interested readers can also refer to [57],[58],[76] and [80].

Figure 4.6(a) is the input image in grayscale. The left side of the image is dark while the right side is light. While traversing from left to right, at the junction between the two regions, the pixel intensity changes from dark to light. Figure 4.6(b) is the intensity profile across a horizontal cross-section of the input image. Notice that at the point of transition from dark region to light region, there is a change in intensity in the profile. Otherwise, the intensity is constant in the dark and light regions respectively. For clarity, only the region around the point of transition is shown in the intensity profile, first derivative (Figure 4.6(c)), and second derivative (Figure 4.6(d)) profiles. In the transition region, since the intensity profile is increasing the first derivative is positive, while being zero in the dark and white regions. First derivative has a maximum or peak at the edge. Since the first derivative is increasing before the edge, the second derivative is positive before the edge. Likewise, since the first derivative is decreasing after the edge, the second derivative is negative after the edge. Also, second derivative is zero in dark and white regions. At the edge, the second derivative is zero. This phenomenon of the second derivative changing the sign from positive before the edge to negative after the edge or viceversa is known as zero-crossing, as it takes a value of zero at the edge. The input image was simulated on a computer and does not have any noise. However, acquired images will have noise that may affect the detection of zero-crossing. Also, if the intensity changes rapidly in the profile, spurious edges will be detected by the zero-crossing. To prevent the issues due to noise or rapidly changing intensity, the image is pre-processed before application of a second derivative filter.

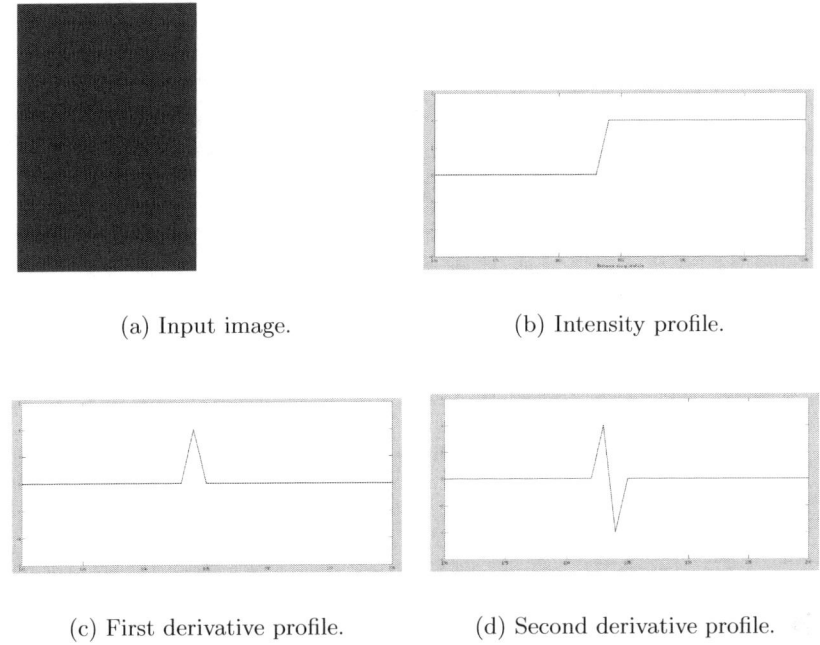

FIGURE 4.6: An example of zero-crossing.

4.3.1 First Derivative Filters

An image is not a continuous function and hence derivatives are calculated using discrete approximations. Let us look at the definition of gradient of a continuous function and then extend it to discrete cases. If $f(x,y)$ is a continuous function, then the gradient of f as a vector is given by

$$\nabla f = \begin{bmatrix} f_x \\ f_y \end{bmatrix} \qquad (4.3)$$

where $f_x = \dfrac{\partial f}{\partial x}$ is known as the partial derivative of f with respect to x, it represents change of f along the horizontal direction and $f_y = \dfrac{\partial f}{\partial y}$ is known as the partial derivative of f with respect to y, it represents change of f along the vertical direction. For more details refer to [86]. The magnitude of the gradient is a scalar quantity and is given by

$$|\nabla f| = [(f_x)^2 + (f_y)^2]^{\frac{1}{2}} \tag{4.4}$$

For computational purposes, we consider the simplified version of the gradient is given by Equation 4.5 and angle is given by Equation 4.6.

$$|\nabla f| = |f_x| + |f_y| \tag{4.5}$$

$$\theta = \tan^{-1}\left(\frac{f_y}{f_x}\right) \tag{4.6}$$

One of the most popular first derivative filters is the Sobel filter. The Sobel filter or mask is used to find horizontal and vertical edges as given in Table 4.3.

-1	-2	-1
0	0	0
1	2	1

-1	0	1
-2	0	2
-1	0	1

TABLE 4.3: Sobel masks for horizontal and vertical edges.

To understand how filtering is done, let us consider a sub-image of size 3-by-3 given in Table 4.4 and multiply the sub-image with horizontal and vertical Sobel masks. The corresponding output is given in Table 4.5.

f_1	f_2	f_3
f_4	f_5	f_6
f_7	f_8	f_9

TABLE 4.4: A 3-by-3 subimage.

Since f_x is the partial derivative of f in x direction which is change of f along horizontal direction, the partial can be obtained by taking the difference between the third row and the first row in the horizontal

$-f_1$	$-2f_2$	$-f_3$
0	0	0
f_7	$2f_8$	f_9

$-f_1$	0	f_3
$-2f_4$	0	$2f_6$
$-f_7$	0	f_8

TABLE 4.5: Output after multiplying the sub-image with Sobel masks.

mask, so $f_x = (f_7 + 2f_8 + f_9) - (-f_1 - 2f_2 - f_3)$. Likewise, f_y is the partial of f in y direction which is change of f in vertical direction, the partial can be obtained by taking the difference between the third column and the first column in the vertical mask, so $f_y = (f_3 + 2f_6 + f_9) - (-f_1 + 2f_4 - f_7)$. Simplying f_x and f_y the discrete gradient at f_5 is given by the Equation 4.7.

$$|f_5| = |f_7 + 2f_8 + f_9 + f_1 + 2f_2 + f_3| + |f_3 + 2f_6 + f_9 + f_1 - 2f_4 + f_7| \quad (4.7)$$

The important features of the Sobel filter are:

- The sum of the coefficients in the mask image is 0. This means that the pixels with constant grayscale are not affected by the derivative filter.

- The side effect of derivative filters is creation of additional noise. Hence, coefficients of $+2$ and -2 are used in the mask image to produce smoothing.

Another popular first derivative filter is Prewitt [77]. The masks for the Prewitt filter are given in Table 4.6.

-1	-1	-1
0	0	0
1	1	1

-1	0	1
-1	0	1
-1	0	1

TABLE 4.6: Prewitt masks for horizontal and vertical edges.

As in the case of Sobel, the sum of the coefficients in Prewitt is also 0. Hence this filter does not affect pixels with constant grayscale.

However, the filter does not reduce noise as can be seen in the values of the coefficients.

The following is the Python function for Sobel filter:

```
filter.sobel(image, mask=None)

Necessary arguments:
  image is an ndarray.

Optional arguments:
  mask is a boolean array. mask is used to
  perform the sobel filter on specific region in
  the image.

Returns: output is an ndarrray.
```

The Python code for the Sobel filter is given below.

```
import scipy.misc
from skimage import filter
from scipy.misc.pilutil import Image

# opening the image and converting it to grayscale
a = Image.open('../Figures/cir.png').convert('L')
# performing Sobel filter
b = filter.sobel(a)
# b is converted from an ndarray to an image
b = scipy.misc.toimage(b)
b.save('../Figures/sobel_cir.png')
```

For Prewitt, the Python function is given below and the arguments are the same as Sobel function arguments:

```
filter.prewitt(image, mask=None)
```

Let us consider an example to illustrate the effect of filtering an image using both Sobel and Prewitt. The image in Figure 4.7(a) is a CT slice of a human skull near the nasal area. The output of Sobel and Prewitt filters is given in Figures 4.7(b) and 4.7(c). Both filters have successfully created the edge image.

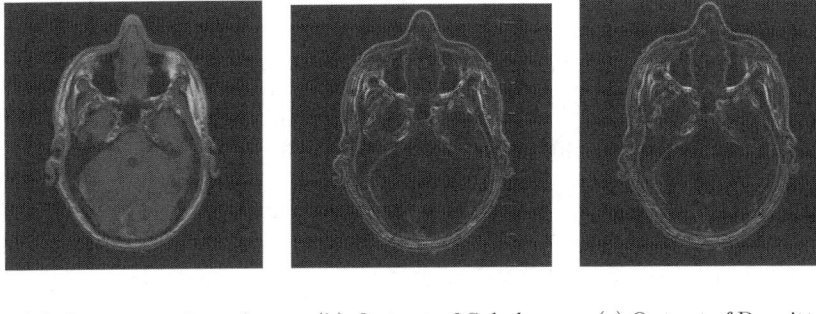

(a) A cross-section of human skull.   (b) Output of Sobel.   (c) Output of Prewitt.

FIGURE 4.7: Example for Sobel and Prewitt.

Slightly modified Sobel and Prewitt filters can be used to detect one or more types of edges. Sobel and Prewitt filters to detect diagonal edges are given in Tables 4.7 and 4.8.

0	1	2
-1	0	1
-2	-1	0

-2	-1	0
-1	0	1
0	1	2

TABLE 4.7: Sobel masks for diagonal edges.

The Python functions to detect vertical and horizontal edges are vsobel, hsobel, vprewitt and hprewitt. The function arguments for all these filters are the same as the function arguments for Sobel discussed

0	1	1
-1	0	1
-1	-1	0

-1	-1	0
-1	0	1
0	1	1

TABLE 4.8: Prewitt masks for diagonal edges.

earlier. For example, for horizontal edge detection, use hprewitt and the Python function is:

```
filter.hprewitt(image,mask = None)
```

Figure 4.8 is an example of detection of horizontal and vertical edges using Sobel and Prewitt. The vertical Sobel and Prewitt filters have enhanced all the vertical edges while the corresponding horizontal filters enhanced horizontal edges and the regular Sobel and Prewitt filters enhanced all edges.

Another popular filter for edge detection is the Canny filter or Canny edge detector [12]. This filter uses three parameters to detect edges. The first parameter is the standard deviation, σ for Gaussian filter. The second and third parameters are the threshold values, t_1 and t_2. The Canny filter can be best described by the following steps:

1. The input of the Canny filter is a grayscale image.

2. A Gaussian filter is used on the image for smoothing.

3. An important property of an edge pixel is that it will have a maximum gradient magnitude in the direction of the gradient. So, for each pixel, the magnitude of the gradient given in Equation 4.5 and the the corresponding direction, $\theta = \tan^{-1}\left(\frac{f_x}{f_y}\right)$ are computed.

4. At the edge points, the first derivative will have either a minimum or a maximum. This implies that the magnitude (absolute value)

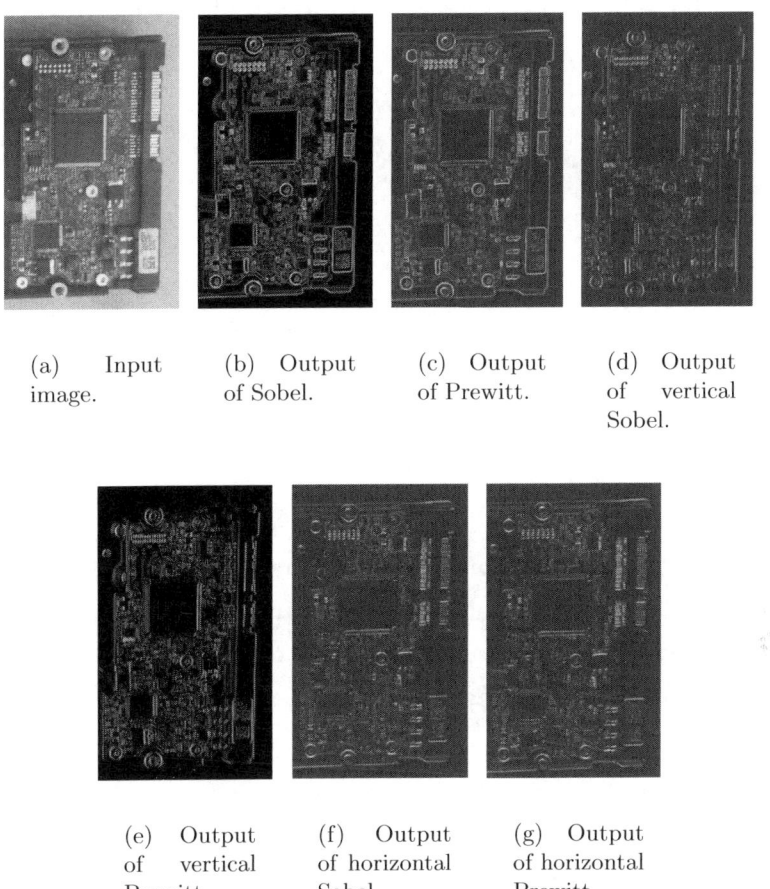

(a) Input image. (b) Output of Sobel. (c) Output of Prewitt. (d) Output of vertical Sobel.

(e) Output of vertical Prewitt. (f) Output of horizontal Sobel. (g) Output of horizontal Prewitt.

FIGURE 4.8: Output from vertical, horizontal and regular Sobel and Prewitt filters.

of gradient of the image at the edge points is maximum. We will refer to these points as ridge pixels. To identify edge points and suppress others, only ridge tops are retained and other pixels are assigned a value of zero. This process is known as non-maximal suppression.

5. Two thresholds, low threshold and high threshold, are then used to threshold the ridges. Ridge pixel values help to classify edge pixels into weak and strong. Ridge pixels with values greater than

the high threshold are classified as strong edge pixels, whereas the ridge pixels between low threshold and high threshold are called weak edge pixels.

6. In the last step, the weak edge pixels are 8-connected with strong edge pixels.

The Python function that is used for the Canny filter is:

```
filter.canny(image, sigma=0.001, low_threshold,
    high_threshold, mask)
```

Necessary arguments:
  image is the input image as an array with floating
  values. The input image should be grayscale image
  normalized to values between 0.0 and 1.0, this can
  be achieved by dividing all the pixel values by the
  maximum pixel value.

Optional arguments:
  sigma is the standard deviation of the Gaussian;
  it is a float. Default value of sigma is 1.0.

  low_threshold is a minimum bound used to connect
  edges; it is a float. Default value is 0.1.

  high_threshold is a maximum bound used to connect
  edges; it is a float. Default value is 0.2.

  mask is a boolean array. mask is used to
  perform the canny filter on specific region in
  the image.

Returns: output is an ndarray.

The Python code for the Canny filter is given below.

```
import scipy.misc, numpy
from skimage import filter
from scipy.misc.pilutil import Image

# opening the image and converting it to grayscale
a = Image.open('../Figures/maps1.png').convert('L')
# converting a to an ndarray
a = scipy.misc.fromimage(a)
# performing Canny edge filter
b = filter.canny(a, sigma=1.0)
# b is converted from ndarray to an image
b = scipy.misc.toimage(b)
# saving b as canny_output.png
b.save('../Figures/canny_output.png')
```

Figure 4.9(a) is a simulated map consisting of names of geographical features of Antarctica. The Canny edge filter is used on this input image to obtain only edges of the letters as shown in Figure 4.9(b).

4.3.2 Second Derivative Filters

In this filter, the second derivative is computed in order to determine the edges. One of the most popular second derivative filters is the Laplacian. The Laplacian of a continuous function is given by:

$$\nabla^2 f = \frac{\partial^2 f}{\partial x^2} + \frac{\partial^2 f}{\partial y^2}$$

where $\frac{\partial^2 f}{\partial x^2}$ is the second partial derivative of f in x direction represents

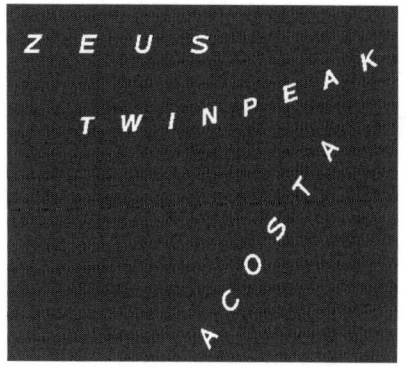

(a) Input image for Canny filter.   (b) Output of Canny filter.

FIGURE 4.9: Example of Canny filter.

change of $\frac{\partial f}{\partial x}$ along the horizontal direction and $\frac{\partial^2 f}{\partial y^2}$ is the second partial derivative of f in y direction represents change of $\frac{\partial f}{\partial y}$ along the vertical direction. For more details refer to [23] and [26]. The discrete Laplacian used for image processing has several versions. Most widely used Laplacian masks are given in Table 4.9.

0	1	0
-1	4	-1
0	-1	0

-1	-1	-1
-1	8	1
-1	-1	-1

TABLE 4.9: Laplacian masks.

The Python function that is used for the Laplacian along with the arguments is the following:

```
scipy.ndimage.filters.laplace(input, output=None,
   mode='reflect', cval=0.0)
```

Necessary arguments:

Spatial Filters

input is the input image as an ndarray

Optional arguments:
 mode determines the method for handling the array border by padding. Different options are: constant, reflect, nearest, mirror, wrap.

 cval is a scalar value specified when the option for mode is constant. The default value is 0.0.

 origin is a scalar that determines origin of the filter. The default value 0 corresponds to a filter whose origin (reference pixel) is at the center. In a 2D case, origin = 0 would mean (0,0).

Returns: output is an ndarray

The Python code for the Laplacian filter is given below. The Laplacian is called using the Python's Laplace function along with the optional mode for handling array borders.

```
import scipy.misc
import scipy.ndimage
from scipy.misc.pilutil import Image

# opening the image and converting it to grayscale
a =Image.open('../Figures/imagefor_laplacian.png').
    convert('L')
# performing Laplacian filter
b = scipy.ndimage.filters.laplace(a,mode='reflect')
# b is converted from an ndarray to an image
b = scipy.misc.toimage(b)
b.save('../Figures/laplacian_new.png')
```

The black-and-white image in Figure 4.10(a) is a segmented CT slice of a human body across the rib cage. The various blobs in the image are the ribs. The Laplacian filter obtained the edges without any artifact.

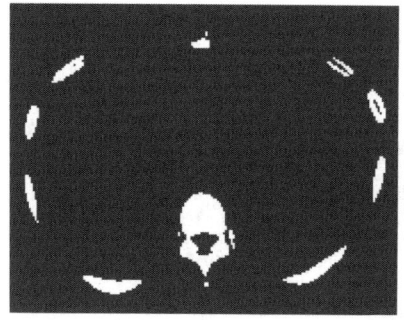 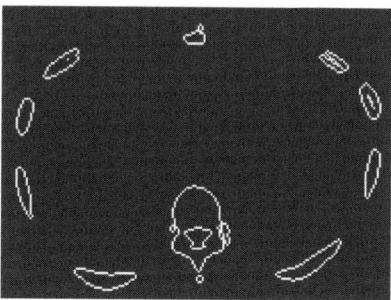

(a) Input image for Laplacian (b) Output of Laplacian

FIGURE 4.10: Example of the Laplacian filter.

As discussed earlier, a derivative filter adds noise to an image. The effect is magnified when the first derivative image is differentiated again (to obtain a second derivative) as in the case of second derivative filters. Figure 4.11 displays this effect. The image in the Figure 4.11(a) is an MRI image of a skull. As there are several edges in the input image, the Laplacian filter over-segments the object (creates many edges) as seen in the output, Figure 4.11(b). This results in a noisy image with no discernable edges.

To offset the noise effect from the Laplacian, a smoothing function, Gaussian, is used along with the Laplacian. While the Laplacian calculates the zero-crossing and determines the edges, the Gaussian smooths the noise induced by the second derivative.

The Gaussian function is given by

$$G(r) = -e^{\frac{-r^2}{2\sigma^2}} \tag{4.8}$$

Spatial Filters

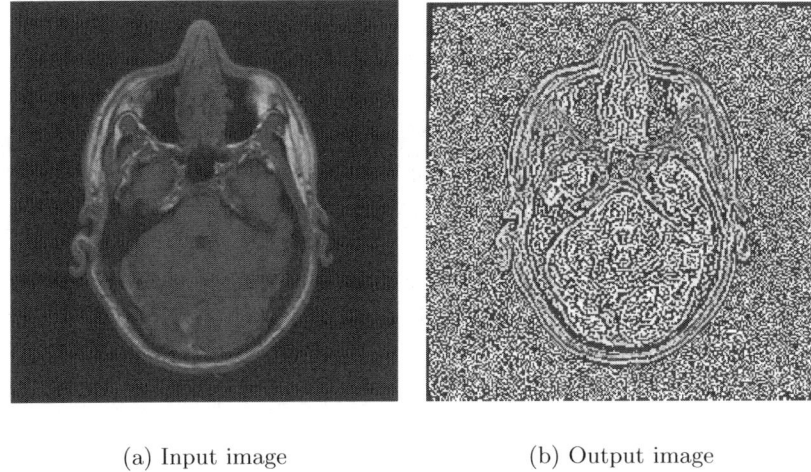

(a) Input image   (b) Output image

FIGURE 4.11: Another example of Laplacian filter.

where $r^2 = x^2 + y^2$ and σ is the standard deviation. A convolution of an image with the Gaussian will result in smoothing of the image. The σ determines the magnitude of smoothing.

The Laplacian convolved with Gaussian is known as the Laplacian of Gaussian and is denoted by LoG. Since Laplacian is the second derivative, the LoG expression can be obtained by finding the second derivative of G with respect to r which yields

$$\nabla^2 G(r) = -\left(\frac{r^2 - \sigma^2}{\sigma^4}\right) e^{-\frac{r^2}{2\sigma^2}} \quad (4.9)$$

The LoG mask or filter of size 5-by-5 is given in Table 4.10.

0	0	-1	0	0
0	-1	-2	-1	0
-1	-2	16	-2	-1
0	-1	-2	-1	0
0	0	-1	0	0

TABLE 4.10: Laplacian of Gaussian mask

The following is the Python function for LoG:

```
scipy.ndimage.filters.gaussian_laplace(input,
    sigma, output=None, mode='reflect', cval=0.0)
```

```
Necessary arguments:
  input is the input image as an ndarray

  sigma is the standard deviation of the Gaussian;
  it is a float
```

```
Optional arguments:
  mode determines the method for handling the array
border by padding. Different options are: constant,
reflect, nearest, mirror, wrap.

  cval is a scalar value specified when the option for
  mode is constant. The default value is 0.0.

  origin is a scalar that determines origin of the
  filter. The default value 0 corresponds to a filter
  whose origin (reference pixel) is at the center. In a
  2D case, origin = 0 would mean (0,0).
```

```
Returns: output is an ndarray
```

The Python code below shows the implementation of the LoG filter. The filter is invoked using gaussian_laplace function with a sigma of 1.

```python
import scipy.misc
import scipy.ndimage
from scipy.misc.pilutil import Image
```

Spatial Filters 85

```
# opening the image and converting it to grayscale
a = Image.open('../Figures/vhuman_t1.png').
    convert('L')
# performing Laplacian of Gaussian
b = scipy.ndimage.filters.gaussian_laplace(a,1,
    mode='reflect')
# b is converted from an ndarray to an image
b = scipy.misc.toimage(b)
b.save('../Figures/log_vh1.png')
```

Figure 4.12(a) is the input image and Figure 4.12(b) is the output after the application of LoG. The LoG filter was able to determine the edges more accurately compared to Laplacian alone. However, the non-uniform foreground intensity has contributed towards formation of blobs (a group of connected pixels).

The major disadvantage of LoG is the computational price as two operations, Gaussian followed by Laplacian, have to be performed. Even though LoG segments the object from the background, it over-segments the edges within the object causing closed loops (also called the spaghetti effect) as shown in the output Figure 4.12(b).

4.4 Summary

- The mean filter smoothens the image while blurring the edges in the image.

- The median filter is effective in removing salt and pepper noise.

- The most widely used first derivative filters are Sobel, Prewitt and Canny.

- Both Laplacian and LoG are popular second derivative filters.

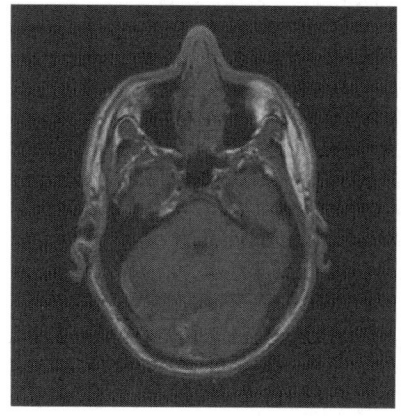 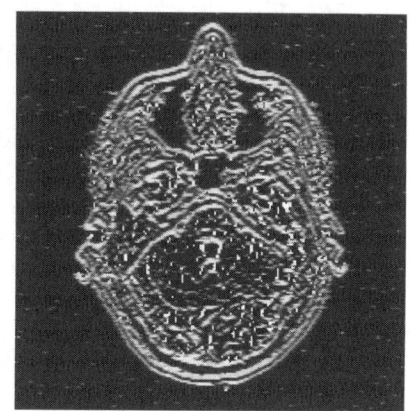

(a) Input image for LoG (b) Output of LoG filter

FIGURE 4.12: Example of LoG.

Laplacian is very sensitive to noise. In LoG, the Gaussian smooths the image so that the noise from the Laplacian can be compensated. But LoG suffers from the spaghetti effect.

4.5 Exercises

1. Write a Python program to apply a mean filter on an image with salt-and-pepper noise. Describe the output, including the mean filter's ability to remove the noise.

2. Describe how effective the mean filter is in removing salt-and-pepper noise. Based on your understanding of the median filter, can you explain why the mean filter cannot remove salt-and-pepper noise?

3. Can max filter or min filter be used for removing salt-and-pepper noise?

4. Check the scipy documentation available at http://docs.scipy.org/doc/scipy/reference/ndimage.html. Identify the Python function that can be used for creating custom filters.

5. Write a Python program to obtain the difference of the Laplacian of Gaussian (LoG). The pseudo code for the program will be as follows:

    (a) Read the image
    (b) Apply LoG assuming a standard deviation of 0.1 and store the image as im1
    (c) Apply LoG assuming a standard deviation of 0.2 and store the image as im2
    (d) Find the difference between the two images and store the resulting image

6. In this chapter, we have discussed a few spatial filters. Identify two more filters and discuss their properties.

Chapter 5

Image Enhancement

5.1 Introduction

In previous chapters we discussed image filters. The filter enhances the quality of image so that important details can be visualized and quantified. In this chapter, we discuss a few more image enhancement techniques. These techniques transform the pixel values in the input image to a new value in the output image using a mapping function. We discuss logarithmic transformation, power law transformation, image inverse, histogram equalization, and contrast stretching. For more information on image enhancement refer to [37],[71],[74].

5.2 Pixel Transformation

A transformation is a function that maps set of inputs to set of outputs so that each input has has exactly one output. For example, $T(x) = x^2$ is a transformation that maps inputs to corresponding squares of input. Figure 5.1 illustrates the transformation $T(x) = x^2$ is given.

In the case of images, a transformation takes the pixel intensities of the image as an input and creates a new image where the corresponding pixel intensities are defined by the transformation. Let us consider $T(x) = x + 50$ transformation. When this transformation is applied to

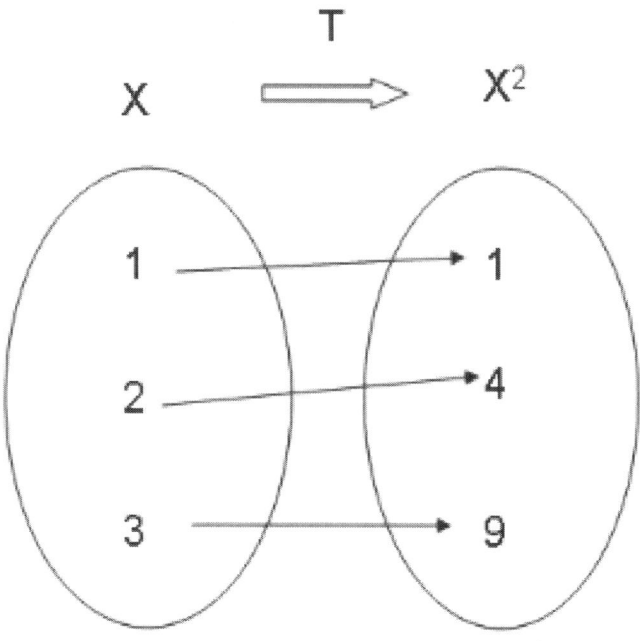

FIGURE 5.1: Illustration of transformation $T(x) = x^2$.

an image, a value of 50 is added to the intensity of each pixel. The corresponding image is brighter than the input image. Figures 5.2(a) and 5.2(b) are the input and output images of the transformation, $T(x) = x + 50$. After the transformation, if the pixel intensity is above $L - 1$, then the intensity of $L - 1$ is assigned to the pixel.

For a grayscale image, the transformation range is given by $[0, L-1]$ where $L = 2^k$ and k is the number of bits in an image. In the case of an 8-bit image, the range is $[0, 2^8 - 1] = [0, 255]$ and for a 16-bit image the range is $[0, 2^{16} - 1] = [0, 65535]$. In this chapter we consider 8-bit grayscale images but the basic principles apply to images of any bit depth.

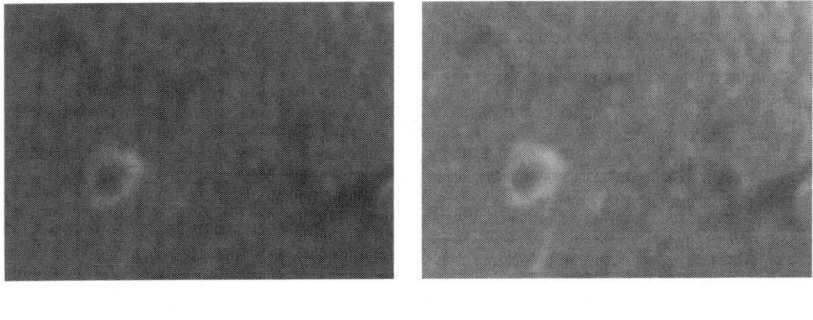

(a) Input image (b) Output image

FIGURE 5.2: Example of transformation $T(x) = x+50$. Original image reprinted with permission from Mr. Karthik Bharathwaj.

5.3 Image Inverse

Image inverse transformation is a linear transformation. The goal is to transform the dark intensities in the input image to bright intensities in the output image and vice versa. If the range of intensities is $[0, L-1]$ for the input image, then the image inverse transformation at (i, j) is given by the following

$$t(i,j) = L - 1 - I(i,j) \qquad (5.1)$$

where I is the intensity value of the pixel in the input image at (i, j).

For an 8-bit image, the Python code for the image inverse is given below:

```
import math
import scipy.misc
import numpy as np
from scipy.misc.pilutil import Image

# opening the image and converting it to grayscale
```

```
im = Image.
    open('../Figures/imageinverse_input.png').
    convert('L')
# im is converted to an ndarray
im1 = scipy.misc.fromimage(im)
# performing the inversion operation
im2 = 255 - im1
# im2 is converted from an ndarray to an image
im3 = scipy.misc.toimage(im2)
# saving the image as imageinverse_output.png in
# Figures folder
im3.save('../Figures/imageinverse_output.png')
```

Figure 5.3(a) is a CT image of the region around the heart. Notice that there are several metal objects, bright spots with streaks, emanating in the image. The bright circular object near the bottom edge is a rod placed in the spine, while two arch shaped metal objects are the valves in the heart. The metal objects are very bright and prevent us from observing other details. The image inverse transformation suppresses the metal objects while enhancing other features of interest such as blood vessels, as shown in Figure 5.3(b).

5.4 Power Law Transformation

Power law transformation, also known as gamma-correction, is used to enhance the quality of the image. The power transformation at (i, j) is given by

$$t(i,j) = k\,I(i,j)^\gamma \qquad (5.2)$$

Image Enhancement 93

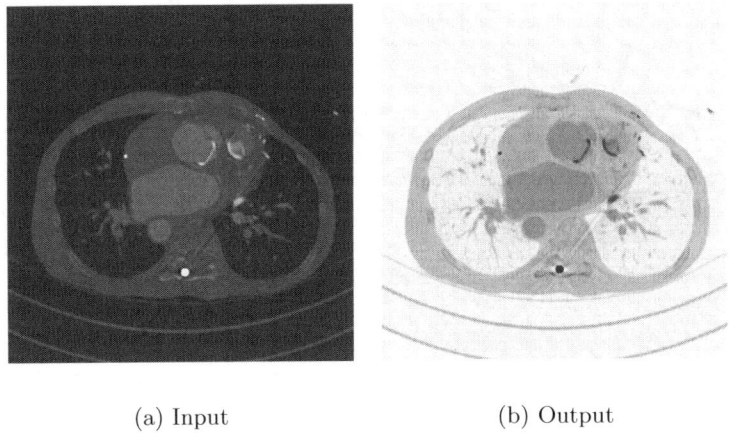

(a) Input (b) Output

FIGURE 5.3: Example of image inverse transformation. Original image reprinted with permission from Dr. Uma Valeti, Cardiovascular Imaging, University of Minnesota.

where k and γ are positive constants and I is the intensity value of the pixel in the input image at (i,j). In most cases $k = 1$.

If $\gamma = 1$ (Figure 5.4), then the mapping is linear and the output image is the same as the input image. When $\gamma < 1$, a narrow range of dark or low intensity pixel values in the input image get mapped to a wide range of intensities in the output image, while a wide range of bright or high intensity pixel values in the input image get mapped to a narrow range of high intensities in the output image. The effect from values of $\gamma > 1$ is opposite that of values $\gamma < 1$. Considering that the intensity range is between $[0,1]$, Figure 5.4 illustrates the effect of different values of γ for $k = 1$.

The human brain uses gamma-correction to process an image, hence gamma-correction is a built-in feature in devices that display, acquire, or publish images. Computer monitors and television screens have built-in gamma-correction so that the best image contrast is displayed in all the images.

In an 8-bit image, the intensity values range from $[0, 255]$. If the transformation is applied according to Equation 5.2, and for $\gamma > 1$

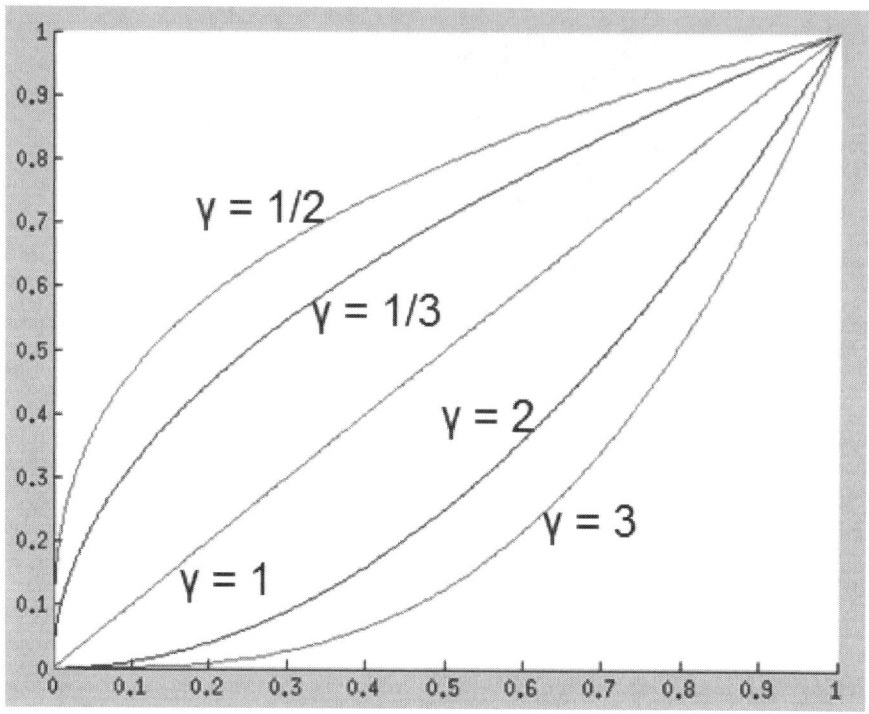

FIGURE 5.4: Graph of power law transformation for different γ.

the output pixel intensities will be out of bounds. To avoid this scenario, in the following Python code the pixel intensities are normalized, $\frac{I(i,j)}{max(I)} = I_{norm}$. For $k = 1$, replacing $I(i,j)$ with I_{norm} and then applying natural log, ln on both sides of Equation 5.2 will result in

$$\ln(t(i,j)) = \ln(I_{norm})^\gamma = \gamma * \ln(I_{norm}) \qquad (5.3)$$

now basing both sides by e will give us

$$e^{\ln(t(i,j))} = e^{\gamma*\ln(I_{norm})} \qquad (5.4)$$

since $e^{\ln(x)} = x$, the left side in the above equation will simplify to

$$t(i,j) = e^{\gamma*\ln(I_{norm})} \qquad (5.5)$$

to have the output in the range $[0, 255]$ we multiply the right side of the above equation by 255 which results in

$$t(i,j) = e^{\gamma*\ln(I_{norm})} * 255. \tag{5.6}$$

This transformation is used in the Python code for power law transformation given below.

```
import math, numpy
import scipy.misc
from scipy.misc.pilutil import Image

# opening the image and converting it to grayscale
a = Image.open('../Figures/angiogram1.png').
    convert('L')
# a is converted to an ndarray
b = scipy.misc.fromimage(a)
# gamma is initialized
gamma = 0.5
# b is converted to type float
b1 = b.astype(float)
# maximum value in b1 is determined
b3 = numpy.max(b1)
# b1 is normalized
b2 = b1/b3
# gamma-correction exponent is computed
b3 = numpy.log(b2)*gamma
# gamma-correction is performed
c = numpy.exp(b3)*255.0
# c is converted to type int
c1 = c.astype(int)
# c1 is converted from ndarray to image
d = scipy.misc.toimage(c1)
# displaying the image
```

d.show()

Figure 5.5(a) is an image of the angiogram of blood vessels. The image is too bright and it is quite difficult to distinguish the blood vessels from background. Figure 5.5(b) is the image after gamma correction with $\gamma = 0.5$; the image is brighter compared to the original image. Figure 5.5(c) is the image after gamma correction with $\gamma = 5$; this image is darker and the blood vessels are visible.

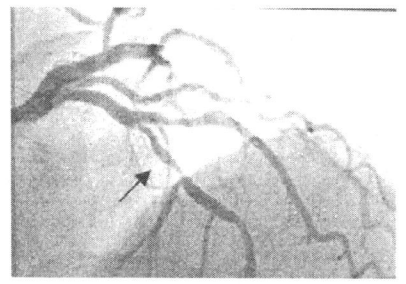

(a) Input image.   (b) Gamma corrected image with $\gamma = 0.5$.

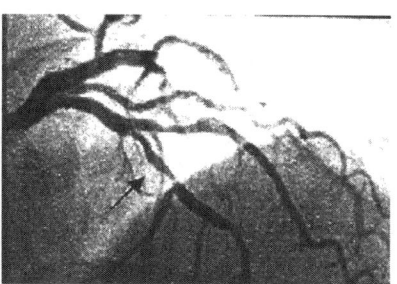

(c) Gamma-corrected image with $\gamma = 5$.

FIGURE 5.5: An example of power law transformation.

5.5 Log Transformation

Log transformation is used to enhance pixel intensities that are otherwise missed due to a wide range of intensity values or lost at the expense of high intensity values. If the intensities in the image range from $[0, L-1]$ then the log transformation at (i,j) is given by

$$t(i,j) = k \log(1 + I(i,j)) \qquad (5.7)$$

where $k = \dfrac{L-1}{\log(1+|I_{max}|)}$ and I_{max} is maximum magnitude value and $I(i,j)$ is the intensity value of the pixel in the input image at (i,j). If both $I(i,j)$ and I_{max} are equal to $L-1$ then $t(i,j) = L-1$. When $I(i,j) = 0$, since $\log(1) = 0$ will give $t(i,j) = 0$. While the end points of the range get mapped to themselves, other input values will be transformed by the above equation. The log can be of any base; however, common log (log base 10) or natural log (*log* base *e*) are widely used. The inverse of the above log transformation when the base is e is given by $t^{-1}(x) = e^{\frac{x}{k}} - 1$ which does the opposite of the log transformation.

Similar to the power law transformation with $\gamma < 1$, the log transformation also maps a small range of dark or low intensity pixel values in the input image to a wide range of intensities in the output image, while a wide range of bright or high intensity pixel values in the input image get mapped to narrow range of high intensities in the output image. Considering the intensity range is between $[0, 1]$, Figure 5.6 illustrates the log and inverse log transformations.

The Python code for log transformation is given below.

```
import scipy.misc
import numpy, math
from scipy.misc.pilutil import Image

# opening the image and converting it to grayscale
```

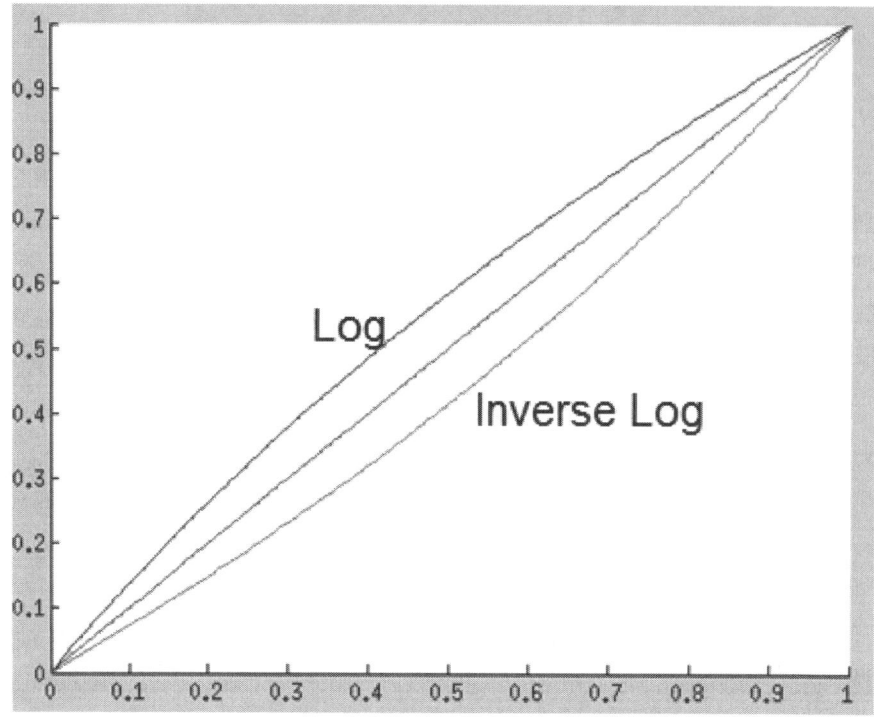

FIGURE 5.6: Graph of log and inverse log transformations.

```
a = Image.open('../Figures/bse.png').convert('L')
# a is converted to an ndarray
b = scipy.misc.fromimage(a)
# b is converted to type float
b1 = b.astype(float)
# maximum value in b1 is determined
b2 = numpy.max(b1)
# performing the log transformation
c = (255.0*numpy.log(1+b1))/numpy.log(1+b2)
# c is converted to type int
c1 = c.astype(int)
# c1 is converted from ndarray to image
d = scipy.misc.toimage(c1)
# saving d as logtransform_output.png
```

```
# in Figures folder
d.save('../Figures/logtransform_output.png')
```

Figure 5.7(a) is a backscattered electron microscope image. Notice that the image is very dark and the details are not clearly visible. Log transformation is performed to improve the contrast, to obtain the output image shown in Figure 5.7(b).

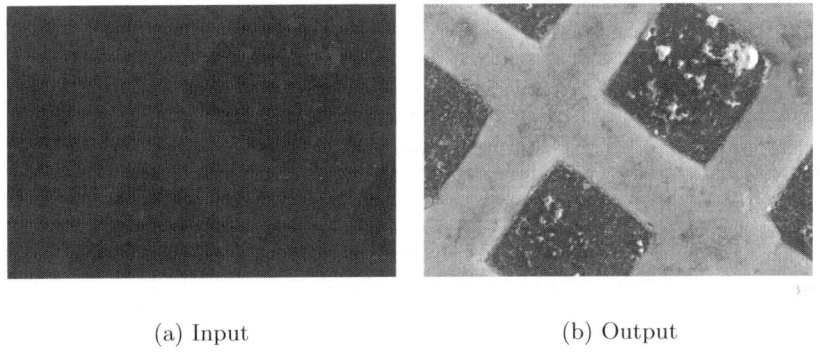

(a) Input (b) Output

FIGURE 5.7: Example of log transformation. Original image reprinted with permission from Mr. Karthik Bharathwaj.

5.6 Histogram Equalization

Image histogram was discussed in Chapter 3, Image and its Properties. The histogram of an image is a discrete function, its input is the gray level value and the output is the number of pixels with that gray level value and can be given as $h(x_n) = y_n$. In a grayscale image, the intensities of the image take values between $[0, L-1]$. As discussed earlier, low gray level values in the image (the left side of the histogram) correspond to dark regions and high gray level values in the image (the right side of the histogram) correspond to bright regions.

In a low contrast image, the histogram is narrow, whereas in a high contrast image, the histogram is spread out. In histogram equalization, the goal is to improve the contrast of an image by rescaling the histogram so that the histogram of the new image is spread out and the pixel intensities range over all possible gray level values. The rescaling of the histogram will be performed by using a transformation. To ensure that for every gray level value in the input image there is a corresponding output, a one-to-one transformation is required; that is every input has a unique output. This means the transformation should be an increasing function. This will ensure that the transformation is invertible.

Before histogram equalization transformation is defined, the following should be computed:

- The histogram of the input image is normalized so that the range of the normalized histogram is $[0, 1]$.

- Since the image is discrete, the probability of a gray level value is denoted by $p_x(i)$ is the ratio of the number of pixels with gray value i to the total number of pixels in the image.

- Cumulative distribution function (CDF) is defined as $C(i) = \sum_{j=0}^{i} p_x(j)$, where $0 \leq i \leq L - 1$ and where L is the total number of gray level values in the image. The $C(i)$ is the sum of all the probabilities of the pixel gray level values from 0 to i. Note that C is an increasing function.

The histogram equalization transformation can be defined as follows:

$$h(u) = round\left(\frac{C(u) - C_{min}}{1 - C_{min}} * (L - 1)\right) \qquad (5.8)$$

where C_{min} is the minimum cumulative distribution value in the image. For a grayscale image with range between $[0, 255]$, if $C(u) = C_{min}$ then

$h(u) = 0$. If $C(u) = 1$ then $h(u) = 255$. The integer value for the output image is obtained by rounding Equation 5.8.

Let us consider an example to illustrate the probability, CDF and histogram equalization. Figure 5.8 is an image of size 5 by 5. Let us assume that the gray levels of the image range from $[0, 255]$.

32	41	52	65	70
60	35	45	66	69
38	65	35	33	39
41	68	75	71	73
37	38	57	58	65

FIGURE 5.8: An example of a 5-by-5 image.

The probabilities, CDF as C for each gray level value for Figure 5.8 along with the output of histogram equalization transformation are given in Figure 5.9.

The Python code for histogram equalization is given below.

```
import numpy as np
import scipy.misc, math
from scipy.misc.pilutil import Image

# opening the image and converting it to grayscale
img = Image.
        open('../Figures/hequalization_input.png').
        convert('L')
```

Gray level value	Probability	CDF as C	h(u)
32	2/25	2/25	11
33	1/25	3/25	22
35	3/25	6/25	54
38	3/25	9/25	85
41	3/25	12/25	117
45	1/25	13/25	128
52	2/25	15/25	149
57	1/25	16/25	160
60	1/25	17/25	170
63	2/25	19/25	192
65	5/25	24/25	245
66	1/25	25/25	255

FIGURE 5.9: Probabilities, CDF, histogram equalization transformation.

```
# img is converted to an ndarray
img1 = scipy.misc.fromimage(img)
# 2D array is convereted to an 1D
fl = img1.flatten()
# histogram and the bins of the image are computed
hist,bins = np.histogram(img1,256,[0,255])
# cumulative distribution function is computed
cdf = hist.cumsum()
# places where cdf=0 is masked or ignored and
# rest is stored in cdf_m
cdf_m = np.ma.masked_equal(cdf,0)
# histogram equalization is performed
num_cdf_m = (cdf_m - cdf_m.min())*255
den_cdf_m = (cdf_m.max()-cdf_m.min())
cdf_m = num_cdf_m/den_cdf_m
```

```
# the masked places in cdf_m are now 0
cdf = np.ma.filled(cdf_m,0).astype('uint8')
# cdf values are assigned in the flattened array
im2 = cdf[fl]
# im2 is 1D so we use reshape command to
#   make it into 2D
im3 = np.reshape(im2,img1.shape)
# converting im3 to an image
im4 = scipy.misc.toimage(im3)
# saving im4 as hequalization_output.png
# in Figures folder
im4.save('../Figures/hequalization_output.png')
```

An example of histogram equalization is illustrated in Figure 5.10. Figure 5.10(a) is a CT scout image. The histogram and CDF of the input image are given in Figure 5.10(b). The output image after histogram equalization is given in Figure 5.10(c). The histogram and cdf of the output image are given in Figure 5.10(d). Notice that the histogram of the input image is narrow compared to the range $[0, 255]$. The leads (bright slender wires running from top to bottom of the image) are not clearly visible in the input image. After histogram equalization, the histogram of the output image is spread out over all the values in the range and subsequently the image is brighter and the leads are visible.

5.7 Contrast Stretching

Contrast stretching is similar in idea to histogram equalization except that the pixel intensities are rescaled using the pixel values instead of probabilities and cdf. Contrast stretching is used to increase the pixel value range by rescaling the pixel values in the input image. Consider an

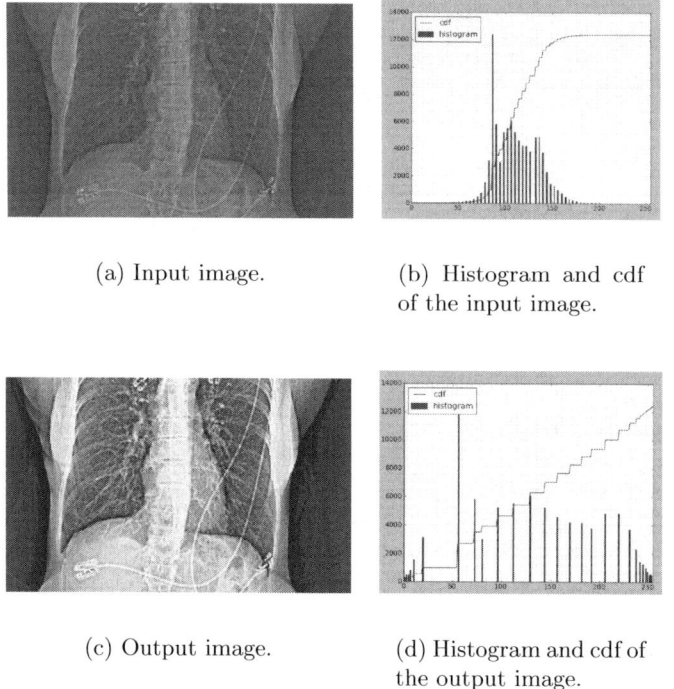

(a) Input image.

(b) Histogram and cdf of the input image.

(c) Output image.

(d) Histogram and cdf of the output image.

FIGURE 5.10: Example of histogram equalization. Original image reprinted with permission from Dr. Uma Valeti, Cardiovascular Imaging, University of Minnesota.

8-bit image with a pixel value range of $[a, b]$ where $a > 0$ and $b < 255$. If a is significantly greater than zero and or if b is significantly smaller than 255, then the details in the image may not be visible. This problem can be offset by rescaling the pixel value range to $[0, 255]$, a much larger pixel range.

The contrast stretching transformation, $t(i,j)$ is given by the following equation:

$$t(i,j) = 255 * \frac{I(i,j) - a}{b - a} \tag{5.9}$$

where $I(i,j)$, a, and b are the pixel intensity at (i,j), the minimum pixel value and the maximum pixel value in the input image respectively.

Image Enhancement

Note that if $a = 0$ and $b = 255$ then there will be no change in pixel intensities between the input and the output images.

```
import math, numpy
import scipy.misc
from scipy.misc.pilutil import Image

# opening the image and converting it to grayscale
im = Image.
    open('../Figures/hequalization_input.png').
    convert('L')
# im is converted to an ndarray
im1 = scipy.misc.fromimage(im)
# finding the maximum and minimum pixel values
b = im1.max()
a = im1.min()
print a,b
# converting im1 to float
c = im1.astype(float)
# contrast stretching transformation
im2 = 255*(c-a)/(b-a)
# im2 is converted from an ndarray to an image
im3 = scipy.misc.toimage(im2)
# saving im3 as contrast_output.png in
# Figures folder
im3.save('../Figures/contrast_output2.png')
```

In Figure 5.11(a) the minimum pixel value in the image is 7 and the maximum pixel value is 51. After contrast strectching, the output image (Figure 5.11(b)) is brighter and the details are visible.

In Figure 5.12(a), the minimum pixel value in the image is equal to 0 and the maximum pixel value is equal to 255 so the contrast stretching

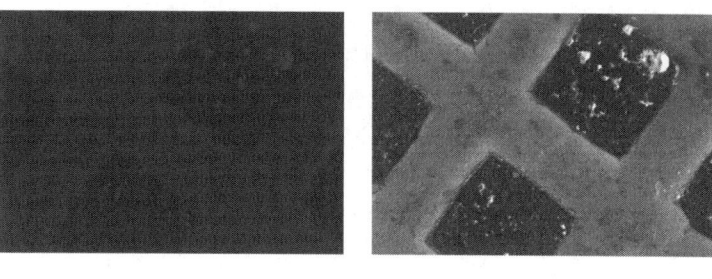

(a) Input image.   (b) Output image.

FIGURE 5.11: An example of contrast stretching where the pixel value range is significantly different from $[0, 255]$.

transformation will not have any effect on this image as shown in Figure 5.12(b).

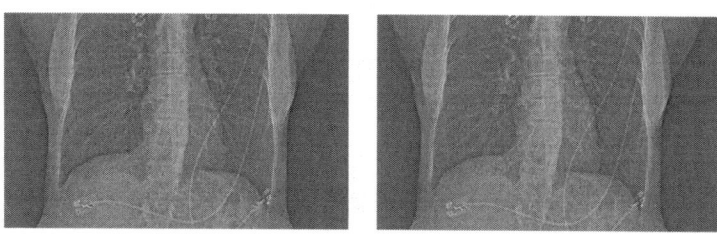

(a) Input image.   (b) Output image.

FIGURE 5.12: An example of contrast stretching where the input pixel value range is same as $[0, 255]$.

5.8 Summary

- Image inverse transformation is used to invert the pixel intensities in an image. This process is similar to obtaining a negative of a photograph.

- Power law transformation makes the image brighter for $\gamma < 1$ and darker for $\gamma > 1$.

- Log transformation makes the image brighter, while the inverse log makes the image darker.

- Histogram equalization is used to enhance the contrast in an image. In this transformation, a narrow range of intensity values will get mapped to a wide range of intensity values.

- Contrast stretching is used to increase the pixel value range by rescaling the pixel values in the input image.

5.9 Exercises

1. Explain briefly the need for image enhancement with some examples.

2. Research a few other image enhancement techniques.

3. Consider an image transformation where every pixel value is multiplied by a constant (K). What will be the effect on the image assuming $K < 1$, $K = 1$ and $K > 1$?

4. All the transformations discussed in this chapter are scaled from $[0, 1]$. Why?

5. The window or level operation allows us to modify the image, so that all pixel values can be visualized. What is the difference between window or level and image enhancement?

    Clue: One makes a permanent change to the image while the other does not.

6. An image has all its pixel values clustered in the lower intensity.

The image needs to be enhanced, so that the small range of the low intensity maps to a larger range. What operation would you use?

Chapter 6

Fourier Transform

6.1 Introduction

In the previous chapters, we focused on images in spatial domain, i.e., the physical world. In this chapter, we will learn about the frequency domain. The process of converting an image from spatial domain to frequency domain provides valuable insight into the nature of the image. In some cases, an operation can be performed more efficiently in the frequency domain than in spatial domain. We introduce the various aspects of Fourier transform and its properties. We focus exclusively on filtering an image in the frequency domain. Interested readers can refer to [7],[97],[102] etc. for more in-depth treatment of Fourier transformation.

The French mathematician Jean Joseph Fourier developed Fourier transforms in an attempt to solve the heat equation. During the process, he recognized that a periodic function can be expressed as infinite sums of sines and cosines of different frequencies, now known as the Fourier series. Fourier transform is an extension of the Fourier series to non-periodic functions. Fourier transform is a representation in which any function can be expressed as the integral of sines and cosines multiplied with the weighted function. Also, any function represented in either Fourier series or transform can be reconstructed completely by an inverse process. This is known as inverse Fourier transform.

This result was published in 1822 in the book "La Theorie Analitique de la Chaleur." This idea was not welcomed, as at that time mathematicians were interested in and studied regular functions. It took over a century to recognize the importance and power of Fourier series and transforms. After the development of the fast Fourier transform algorithm, FFT, the applications of Fourier transforms have affected several fields, remote sensing, signal processing and image processing.

In image processing, Fourier transforms are used for:

- Image filtering
- Image compression
- Image enhancement
- Image restoration
- Image analysis
- Image reconstruction

In this chapter we discuss image filtering and enhancement in detail.

6.2 Definition of Fourier Transform

A Fourier transform of a continuous function in one variable $f(x)$ is given by the following equation:

$$F(u) = \int_{-\infty}^{\infty} f(x) e^{-i2\pi ux} dx \qquad (6.1)$$

where $i = \sqrt{-1}$. The function $f(x)$ can be retrieved by finding the inverse Fourier transform of $F(u)$ which is given by the following equation:

$$f(x) = \int_{-\infty}^{\infty} F(u) e^{i2\pi ux} du. \qquad (6.2)$$

Fourier Transform

The Fourier transform of a one variable discrete function, $f(x)$ for $x = 0, 1, ... L - 1$ is given by the following equation:

$$F(u) = \frac{1}{L} \sum_{x=0}^{L-1} f(x) e^{\frac{-i2\pi ux}{L}} \tag{6.3}$$

for $u = 0, 1, 2, ..., L - 1$. Equation 6.3 is known as the discrete Fourier transform, DFT. Likewise, the inverse discrete Fourier transform, (IDFT) is given by the following equation:

$$f(x) = \sum_{x=0}^{L-1} F(u) e^{\frac{i2\pi ux}{L}} \tag{6.4}$$

for $x = 0, 1, 2, ..., L - 1$. Using the Euler's formula $e^{i\theta} = \cos\theta + i \sin\theta$, the above equation simplifies to

$$F(u) = \frac{1}{L} \sum_{x=0}^{L-1} f(x) \left[\cos\left(\frac{-2ux\pi}{L}\right) - i \sin\left(\frac{-i2ux\pi}{L}\right) \right] \tag{6.5}$$

Now, using the fact that cos is an even function, i.e., $\cos(-\pi) = \cos(\pi)$ and that sin is an odd function, i.e., $\sin(-\pi) = -\sin(\pi)$, Equation 6.5 can be simplified to:

$$F(u) = \frac{1}{L} \sum_{x=0}^{L-1} f(x) \left[\cos\left(\frac{2ux\pi}{L}\right) + i \sin\left(\frac{2ux\pi}{L}\right) \right] \tag{6.6}$$

$F(u)$ has two parts; the real part constituting cos is represented as $R(u)$ and the imaginary part constituting sin is represented as $I(u)$. Each term of F is known as the coefficient of the Fourier transform. Since u plays a key role in determining the frequency of the coefficients of the Fourier transform, u is known as the frequency variable, while x is known as the spatial variable.

Traditionally many experts have compared the Fourier transform to a glass prism. As a glass prism splits or separates the light into various wavelengths or frequencies that form a spectrum, Fourier transform splits or separates a function into its coefficients which depend on the frequency. These Fourier coefficients form a Fourier spectrum in the frequency domain.

From Equation 6.6, we know that the Fourier transform is comprised of complex numbers. For computational purposes, it is convenient to represent the Fourier tranform in polar form as:

$$F(u) = |F(u)|e^{-i\theta(u)} \tag{6.7}$$

where $|F(u)| = \sqrt{R^2(u) + I^2(u)}$ is called the magnitude of the Fourier transform and $\theta(u) = \tan^{-1}\left[\frac{I(u)}{R(u)}\right]$ is called the phase angle of the transform. Power, $P(u)$, is defined as the following:

$$P(u) = R^2(u) + I^2(u) = |F(u)|^2. \tag{6.8}$$

The first value in the discrete Fourier transform is obtained by setting $u = 0$ in equation (6.3) and then summing the product over all x. Hence, $F(0)$ is nothing but the average of $f(x)$ since $e^0 = 1$. $F(0)$ has the real part while the imaginary part is zero. Other values of F can be computed in a similar manner.

Let us consider a simple example to illustrate the Fourier transform. Let $f(x)$ be a discrete function with only four values: $f(0) = 2, f(1) = 3, f(2) = 2$ and $f(3) = 1$. Note that the size of f is 4, hence $L = 4$.

$$F(0) = \frac{1}{4}\sum_{x=0}^{3} f(x) = \frac{f(0) + f(1) + f(2) + f(3)}{4} = 2$$

$$F(1) = \frac{1}{4}\sum_{x=0}^{3} f(x)\left[\cos\left(\frac{-2\pi x}{4}\right) - i\sin\left(\frac{-i2\pi x}{4}\right)\right]$$

$$= \frac{1}{4}\left(f(0)\left[\cos\left(\frac{0}{4}\right) + i\sin\left(\frac{0}{4}\right)\right] + f(1)\left[\cos\left(\frac{2\pi}{4}\right) + i\sin\left(\frac{2\pi}{4}\right)\right]\right.$$

$$\left. + f(2)\left[\cos\left(\frac{4\pi}{4}\right) + i\sin\left(\frac{4\pi}{4}\right)\right] + f(3)\left[\cos\left(\frac{6\pi}{4}\right) + i\sin\left(\frac{6\pi}{4}\right)\right]\right)$$

$$= \frac{1}{4}(2(1+0i) + 3(0+1i) + 2(-1+0i) + 1(0-1i))$$

$$= \frac{2i}{4} = \frac{i}{2}$$

Note that $F(1)$ is purely imaginary. For $u = 2$, the value of $F(2) = 0$ and for $u = 3$, the value of $F(3) = \frac{-i}{2}$. The four coefficients of the Fourier transform are $\{2, \frac{i}{2}, 0, \frac{-i}{2}\}$.

6.3 Two-Dimensional Fourier Transform

The Fourier transform for two variables is given by the following equation:

$$F(u,v) = \int_{-\infty}^{\infty}\int_{-\infty}^{\infty} f(x,y)\, e^{-i2\pi(ux+vy)}\, dx\, dy \quad (6.9)$$

and the inverse Fourier transform is

$$f(x,y) = \int_{-\infty}^{\infty}\int_{-\infty}^{\infty} F(u,v) e^{i2\pi(ux+vy)}\, du\, dv. \quad (6.10)$$

The discrete Fourier transform of a 2D function, $f(x,y)$ with size L and K is given by the following equation:

$$F(u,v) = \frac{1}{LK}\sum_{x=0}^{L-1}\sum_{y=0}^{K-1} f(x,y) e^{-i2\pi\left(\frac{ux}{L}+\frac{vy}{K}\right)} \quad (6.11)$$

for $u = 1, 2, ..., L-1$ and $v = 1, 2, ..., K-1$. Similar to 1D Fourier transform, $f(x, y)$ can be computed from $F(u, v)$ by computing the inverse Fourier transform, given by the following equation:

$$f(x,y) = \sum_{u=0}^{L-1} \sum_{v=0}^{K-1} F(u,v) e^{i2\pi \left(\frac{ux}{L} + \frac{vy}{K}\right)} \tag{6.12}$$

for $x = 1, 2, ..., L-1$ and $y = 1, 2, ..., K-1$. As in the case of 1D DFT, u and v are the frequency variables and x and y are the spatial variables. The magnitude of the Fourier transform in 2D is given by the following equation:

$$|F(u,v)| = \sqrt{R^2(u,v) + I^2(u,v)} \tag{6.13}$$

and the phase angle is given by

$$\theta(u,v) = \tan^{-1}\left[\frac{I(u,v)}{R(u,v)}\right] \tag{6.14}$$

and the power is given by

$$P(u,v) = R^2(u,v) + I^2(u,v) = |F(u,v)|^2. \tag{6.15}$$

where $R(u, v)$ and $I(u, v)$ are the real and imaginary parts of the 2D DFT.

The properties of a 2D Fourier transform are:

1. The 2D space with x and y as variables is referred to as spatial domain and the space with u and v as variables is referred to as frequency domain.

2. $F(0, 0)$ is the average of all pixel values in the image. It can be obtained by substituting $u = 0$ and $v = 0$ in the equation above. Hence $F(0, 0)$ is the brightest pixel in the Fourier transform image.

3. The two summations are separable. Thus, summation is per-

formed along the x or y-directions first and in the other direction later.

4. The complexity of DFT is N^2. Hence a modified method called Fast Fourier Transform (FFT) is used to calculate the Fourier transform. Cooley and Tukey developed the FFT algorithm [14]. FFT has a complexity of $NlogN$ and hence the word "Fast" in its name.

6.3.1 Fast Fourier Transform using Python

The following is the Python function for the forward Fast Fourier transform:

```
numpy.fft.fft2(a, s=None, axes=(-2,-1))

Necessary arguments:
  a is the input image as an ndarray

Optional arguments:
s is a tuple of integers that represents the
length of each transformed axis of the output.
The individual elements in s, correspond to
the length of each axis in the input image.
If the length on any axis is less than the
corresponding size in the input image, then
the input image along that axis is cropped. If the
length on any axis is greater than the corresponding
size in the input image, then the input image along
that axis is padded with 0s.

  axes is an integer used to compute the FFT. If axis
is not specified, the last two axes are used.
```

Returns: output is a complex ndarray.

The Python code for the forward fast Fourier transform is given below.

```
import math, numpy
import scipy.fftpack as fftim
from scipy.misc.pilutil import Image

# opening the image and converting it to grayscale
a = Image.open('../Figures/fft1.png').convert('L')
# a is converted to an ndarray
b = numpy.asarray(a)
# performing FFT
c = abs(fftim.fft2(b))
# shifting the Fourier frequency image
d = fftim.fftshift(c)
# converting the d to floating type and saving it
# as fft1_output.raw in Figures folder
d.astype('float').
    tofile('../Figures/fft1_output.raw')
```

In the above code, the image data is converted to a numpy array by using the asarray() function. This is similar to the fromimage() function. The Fast Fourier transform is obtained using the fft2 function and only the absolute value is obtained for visualization. The absolute value image of FFT is then shifted, so that the center of the image is the center of the Fourier spectrum. The center pixel corresponds to a frequency of 0 in both directions. Finally, the shifted image is saved as a raw file.

The image in Figure 6.1(a) is a slice of Sindbis virus from a transmission electron microscope. The output after performing the FFT is saved as a raw file since the pixel intensities are floating values. ImageJ is used to obtain the logarithm of the raw image and the window level is adjusted to display the corresponding image. Finally, a spanshot of this image is shown in Figure 6.1(b). As discussed previously, the central pixel is the pixel with the highest intensity. This is due to the fact that the average of all pixel value in the original image consitutes the central pixel. The central pixel is $(0,0)$, the origin. To the left $(0,0)$ is $-u$ and to the right is $+u$. Similarly, to the top of $(0,0)$ is $+v$ and to the bottom is $-v$. The lower frequency is close to the central pixel and the higher frequency is away from the central pixel.

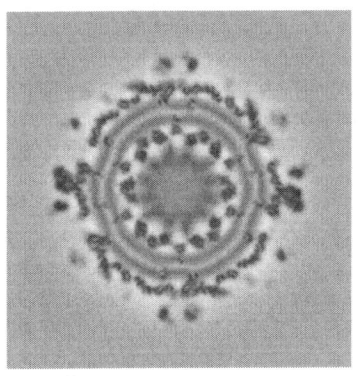
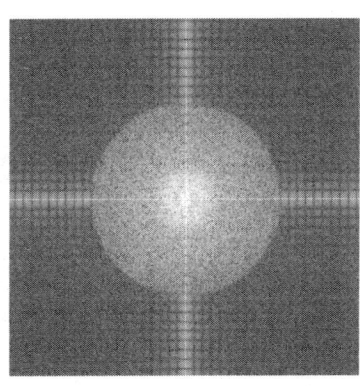

(a) Input for FFT. (b) Output of FFT.

FIGURE 6.1: An example of 2D Fast Fourier transform. Original image reprinted with permission from Dr. Wei Zhang, University of Minnesota.

The Python function for inverse Fast Fourier transform is given below.

```
numpy.fft.ifft2(a, s=None, axes=(-2,-1))
```

```
Necessary arguments:
   a is a complex ndarray comprising of Fourier
transformed data.
```

Optional arguments:

s is a tuple of integers that represents the length of each transformed axis of the output. The individual elements in s, correspond to the length of each axis in the input image. If the length on any axis is less than the corresponding size in the input image, then the input image along that axis is cropped. If the length on any axis is greater than the corresponding size in the input image, then the input image along that axis is padded with 0s.

axes is an integer used to compute the FFT. If axis is not specified, the last two axes are used.

Returns: output is a complex ndarray.

6.4 Convolution

Convolution was briefly discussed in Chapter 4, Spatial Filters, without any mathematical underpinning. In this section, we discuss the mathematical aspects of convolution.

Convolution is a mathematical operation that expresses the integral of the overlap between two functions. A simple example is a blurred image, which is obtained by convolving an un-blurred image with a blurring function.

There are many cases of blurred images that we see in real life. A photograph of a car moving at high speed is blurred due to motion.

A photograph of a star obtained from a telescope is blurred by the particles in the atmosphere. A wide-field microscope image of an object is blurred by a signal from out-of-plane. Such blurring can be modeled as convolution operation and eliminated by the inverse process called deconvolution.

We begin the discussion with convolution in Fourier space. Convolution in spatial domain already has been dealt with in Chapter 4, Spatial Filters. The operation is simpler in Fourier space than in real space but depending on the size of the image and the functions used, the former can be computationally efficient. In Fourier space, convolution is performed on the whole image at once. However, in spatial domain convolution is performed by sliding the filter window on the image.

The convolution operation is expressed mathematically as:

$$[f * g](t) = \int_0^t f(\tau)g(t-\tau)d\tau \qquad (6.16)$$

where f, g are the two functions and the * (asterisk) represents convolution.

The convolution satisfies the following properties:

1. $f * g = g * f$ Commutative Property

2. $f * (g * h) = (f * g) * h$ Assocoative Property

3. $f * (g + h) = f * g + f * h$ Distributive Property

6.4.1 Convolution in Fourier Space

Let us assume that the convolution of f and g is the function h.

$$h(t) = [f * g](t). \qquad (6.17)$$

If the Fourier transform of this function is H, then H is defined as

$$H = F.G \qquad (6.18)$$

where F and G are the Fourier transforms of the functions f and g respectively and the . (dot) represents multiplication. Thus, in Fourier space the complex operation of convolution is replaced by a more simple multiplication. The proof of this theorem is beyond the scope of this book. You can find details in most mathematical textbooks on Fourier transform. The formula is applicable irrespective of the number of dimensions of f and g. Hence it can be applied to a 1D signal and also to 3D volume data.

6.5 Filtering in Frequency Domain

In this section, we discuss applying various filters to an image in the Fourier space. The convolution principle stated in Equation 6.18 will be used for filtering. In lowpass filters, only low frequencies from the Fourier transform are used while high frequencies are blocked. Similarly, in highpass filters, only high frequencies from the Fourier transform are used while the low frequencies are blocked. Lowpass filters are used to smooth the image or reduce noise whereas highpass filters are used to sharpen edges. In each case, three different filters, namely; ideal, Butterworth and Gaussian, are considered. The three filters differ in the creation of the windows used in filtering.

6.5.1 Ideal Lowpass Filter

The convolution function for the 2D ideal lowpass filter (ILPF) is given by

$$H(u,v) = \begin{cases} 1, & \text{if } d(u,v) \leq d_0 \\ 0, & \text{else} \end{cases} \quad (6.19)$$

where d_0 is a specified quantity and $d(u.v)$ is the Euclidean distance from the point (u, v) to the origin of the Fourier domain. Note that for an image of size M by N, the coordinates of the origin are $\left(\dfrac{M}{2}, \dfrac{N}{2}\right)$. So d_0 is the distance of the cutoff frequency from the origin.

For a given image, after the convolution function is defined, the ideal lowpass filter can be performed with element by element multiplication of the FFT of the image and the convolution function. Then the inverse FFT is performed on the convolved function to get the output image.

The Python code for the ideal lowpass filter is given below.

```
import scipy.misc
import numpy, math
import scipy.fftpack as fftim
from scipy.misc.pilutil import Image

# opening the image and converting it to grayscale
a = Image.open('../Figures/fft1.png').convert('L')
# a is converted to an ndarray
b = numpy.asarray(a)
# performing FFT
c = fftim.fft2(b)
# shifting the Fourier frequency image
d = fftim.fftshift(c)

# intializing variables for convolution function
M = d.shape[0]
N = d.shape[1]
# H is defined and
# values in H are initialized to 1
H = numpy.ones((M,N))
center1 = M/2
center2 = N/2
d_0 = 30.0 # cut-off radius
```

```
# defining the convolution function for ILPF
for i in range(1,M):
    for j in range(1,N):
        r1 = (i-center1)**2+(j-center2)**2
        # euclidean distance from
        # origin is computed
        r = math.sqrt(r1)
        # using cut-off radius to eliminate
        # high frequency
        if r > d_0:
            H[i,j] = 0.0
# converting H to an image
H =  scipy.misc.toimage(H)
# performing the convolution
con = d * H
# computing the magnitude of the inverse FFT
e = abs(fftim.ifft2(con))
# e is converted from an ndarray to an image
f =  scipy.misc.toimage(e)
# saving the image as ilowpass_output.png in
# Figures folder
f.save('../Figures/ilowpass_output.png')
```

The image is read and its Fourier transform is determined using the fft2 function. The Fourier spectrum is shifted to the center of the image using the fftshift function. A filter (H) is created by assigning a value of 1 to all pixels within a radius of d_0 and 0 otherwise. Finally, the filter (H) is convolved with the image (d) to obtain the convolved Fourier image (con). This image is inverted using ifft2 to obtain the filtered image in spatial domain. Since high frequencies are blocked, the image is blurred.

A simple image compression technique can be created using the concept of lowpass filtering. In this technique, all high frequency data is cleared and only the low frequency data is stored. This reduces the number of Fourier coefficients stored and consequently needs less storage space on the disk. During the process of displaying the image, an inverse Fourier transform can be obtained to convert the image to spatial domain. Such an image will suffer from blurring, as high frequency information is not stored. A proper selection of the cut-off radius is more important in image compression to avoid blurring and loss of crucial data in the decompressed image.

6.5.2 Butterworth Lowpass Filter

The convolution function for the Butterworth lowpass filter (BLPF) is given below:

$$H(u,v) = \frac{1}{1 + \left(\frac{d(u,v)}{d_0}\right)^2} \quad (6.20)$$

where d_0 is the cut-off distance from the origin for the frequency and $d(u,v)$ is the Euclidean distance from the origin. In this filter, unlike the ILPF, the pixel intensity at the cut-off radius does not change rapidly.

The Python code for the Butterworth lowpass filter is given below:

```
import numpy, math
import scipy.misc
import scipy.fftpack as fftim
from scipy.misc.pilutil import Image


# opening the image and converting it to grayscale
a = Image.open('../Figures/fft1.png').convert('L')
# a is converted to an ndarray
b = scipy.misc.fromimage(a)
# performing FFT
c = fftim.fft2(b)
# shifting the Fourier frequency image
```

```
d = fftim.fftshift(c)


# intializing variables for convolution function
M = d.shape[0]
N = d.shape[1]
# H is defined and
# values in H are initialized to 1
H = numpy.ones((M,N))
center1 = M/2
center2 = N/2
d_0 = 30.0 # cut-off radius
t1 = 1 # the order of BLPF
t2 = 2*t1


# defining the convolution function for BLPF
for i in range(1,M):
    for j in range(1,N):
        r1 = (i-center1)**2+(j-center2)**2
        # euclidean distance from
        # origin is computed
        r = math.sqrt(r1)
        # using cut-off radius to
        # eliminate high frequency
        if r > d_0:
            H[i,j] = 1/(1 + (r/d_0)**t1)

# converting H to an image
H = scipy.misc.toimage(H)
# performing the convolution
con = d * H
# computing the magnitude of the inverse FFT
e = abs(fftim.ifft2(con))
# e is converted from an ndarray to an image
```

```
f = scipy.misc.toimage(e)
# f.show()
# saving the image as blowpass_output.png in
# Figures folder
f.save('../Figures/blowpass_output.png')
```

This program is similar to the Python code used for ILPF except for the creation of the filter (H).

6.5.3 Gaussian Lowpass Filter

The convolution function for the Gaussian lowpass filter (GLPF) is given below:

$$H(u,v) = e^{\frac{-d^2(u,v)}{2d_0^2}} \tag{6.21}$$

where d_0 is the cut-off frequency and $d(u,v)$ is the Euclidean distance from origin. The filter creates a much more gradual change in intensity at the cut-off radius compared to Butterworth lowpass filter.

The Python code for the Gaussian lowpass filter is given below.

```
import numpy, math
import scipy.misc
from scipy.misc import imshow
import scipy.fftpack as fftim
from scipy.misc.pilutil import Image

# opening the image and converting it to grayscale
a = Image.open('../Figures/fft1.png').convert('L')
# a is converted to an ndarray
b = scipy.misc.fromimage(a)
# performing FFT
c = fftim.fft2(b)
# shifting the Fourier frequency image
```

```
d = fftim.fftshift(c)

# intializing variables for convolution function
M = d.shape[0]
N = d.shape[1]
# H is defined and
# values in H are initialized to 1
H = numpy.ones((M,N))
center1 = M/2
center2 = N/2
d_0 = 30.0 # cut-off radius
t1 = 2*d_0


# defining the convolution function for GLPF
for i in range(1,M):
    for j in range(1,N):
        r1 = (i-center1)**2+(j-center2)**2
        # euclidean distance from
        # origin is computed
        r = math.sqrt(r1)
        # using cut-off radius to
        # eliminate high frequency
        if r > d_0:
            H[i,j] = math.exp(-r**2/t1**2)

# converting H to an image
# H = PIL.toimage(H)
H =  scipy.misc.toimage(H)
# performing the convolution
con = d * H
# computing the magnitude of the inverse FFT
e = abs(fftim.ifft2(con))
# e is converted from an ndarray to an image
```

```
f = scipy.misc.toimage(e)
# saving the image as glowpass_output.png in
# Figures folder
f.save('../Figures/glowpass_output.png')
```

Figure 6.1 is the input image to be filtered using ILPF, BLPF and GLPF. The images in Figures 6.2(a), 6.2(b) and 6.2(c) are the outputs of ideal lowpass, Butterworth lowpass, and Gaussian lowpass filters with cut-off radius at 30. Notice how the blurriness varies in the output images. The ILPF is extremely blurred due to the sharp change in the ILPF convolution function at the cut-off radius. There are also severe ringing artifacts, the spaghetti like structure in the background next to the foreground pixels. In BLPF, the convolution function is continuous which results in less blurring and ringing artifacts compared to ILPF. Since a smoothing operator forms the GLPF convolution function, the output of GLPF is even less blurred when compared to both ILPF and BLPF.

6.5.4 Ideal Highpass Filter

The convolution function for the 2D ideal highpass filter (IHPF) is given by

$$H(u,v) = \begin{cases} 0, & \text{if } d(u,v) \leq d_0 \\ 1, & \text{else} \end{cases} \quad (6.22)$$

where d_0 is the cutoff frequency and $d(u,v)$ is the Euclidean distance from the origin.

The Python code for ideal highpass filter is given below.

```
import scipy.misc
import numpy, math
import scipy.fftpack as fftim
from scipy.misc.pilutil import Image
```

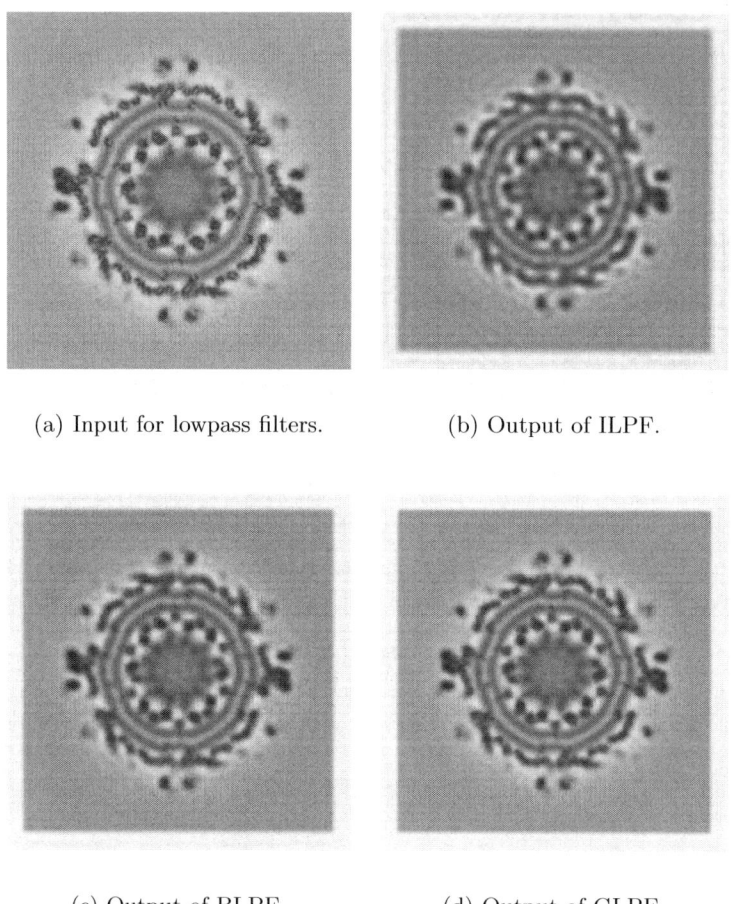

FIGURE 6.2: An example of lowpass filters. The input image and all the output images are displayed in spatial domain.

```
# opening the image and converting it to grayscale
a = Image.open('../Figures/endothelium.png').
    convert('L')
# a is converted to an ndarray
b = scipy.misc.fromimage(a)
# performing FFT
c = fftim.fft2(b)
```

```
# shifting the Fourier frequency image
d = fftim.fftshift(c)


# intializing variables for convolution function
M = d.shape[0]
N = d.shape[1]
# H is defined and
# values in H are initialized to 1
H = numpy.ones((M,N))
center1 = M/2
center2 = N/2
d_0 = 30.0 # cut-off radius


# defining the convolution function for IHPF
for i in range(1,M):
    for j in range(1,N):
        r1 = (i-center1)**2+(j-center2)**2
        # euclidean distance from
# origin is computed
        r = math.sqrt(r1)
        # using cut-off radius to
        # eliminate low frequency
        if 0 < r < d_0:
            H[i,j] = 0.0
# converting H to an image
H = scipy.misc.toimage(H)
# performing the convolution
con = d * H
# computing the magnitude of the inverse FFT
e = abs(fftim.ifft2(con))
# e is converted from an ndarray to an image
f = scipy.misc.toimage(e)
# f.show()
```

```
# saving the image as ihighpass_output.png in
# Figures folder
f.save('../Figures/ihighpass_output.png')
```

In this program, the filter (H) is created by assigning pixel value of 1 to all pixels above the cut-off radius and 0 otherwise.

6.5.5 Butterworth Highpass Filter

The convolution function for the Butterworth highpass filter (BHPF) is given below:

$$H(u,v) = \frac{1}{1 + \left(\frac{d_0}{d(u,v)}\right)^{2n}} \quad (6.23)$$

where d_0 is the cut-off frequency, $d(u,v)$ is the Euclidean distance from origin and n is the order of BHPF.

The Python code for BHPF is given below.

```
import numpy, math
import scipy.misc
import scipy.fftpack as fftim
from scipy.misc.pilutil import Image

# opening the image and converting it to grayscale
a = Image.open('../Figures/endothelium.png').\
    convert('L')
# a is converted to an ndarray
b = scipy.misc.fromimage(a)
# performing FFT
c = fftim.fft2(b)
# shifting the Fourier frequency image
d = fftim.fftshift(c)
```

```python
# intializing variables for convolution function
M = d.shape[0]
N = d.shape[1]
# H is defined and
# values in H are initialized to 1
H = numpy.ones((M,N))
center1 = M/2
center2 = N/2
d_0 = 30.0 # cut-off radius
t1 = 1 # the order of BHPF
t2 = 2*t1


# defining the convolution function for BHPF
for i in range(1,M):
    for j in range(1,N):
        r1 = (i-center1)**2+(j-center2)**2
        # euclidean distance from
        # origin is computed
        r = math.sqrt(r1)
        # using cut-off radius to
        # eliminate low frequency
        if 0 < r < d_0:
            H[i,j] = 1/(1 + (r/d_0)**t2)


# converting H to an image
H = scipy.misc.toimage(H)
# performing the convolution
con = d * H
# computing the magnitude of the inverse FFT
e = abs(fftim.ifft2(con))
# e is converted from an ndarray to an image
f = scipy.misc.toimage(e)
# saving the image as bhighpass_output.png in
```

6.5.6 Gaussian Highpass Filter

The convolution function for the Gaussian highpass filter (GHPF) is given below:

$$H(u,v) = 1 - e^{\frac{-d^2(u,v)}{2d_0^2}} \tag{6.24}$$

where d_0 the cut-off frequency and $d(u,v)$ the Euclidean distance from origin.

The Python code for GHPF is given below.

```python
import numpy, math
import scipy.misc
import scipy.fftpack as fftim
from scipy.misc.pilutil import Image

# opening the image and converting it to grayscale
a = Image.open('../Figures/endothelium.png').\
    convert('L')
# a is converted to an ndarray
b = scipy.misc.fromimage(a)
# performing FFT
c = fftim.fft2(b)
# shifting the Fourier frequency image
d = fftim.fftshift(c)

# intializing variables for convolution function
M = d.shape[0]
N = d.shape[1]
# H is defined and values in H are initialized to 1
```

```
H = numpy.ones((M,N))
center1 = M/2
center2 = N/2
d_0 = 30.0 # cut-off radius
t1 = 2*d_0


# defining the convolution function for GHPF
for i in range(1,M):
    for j in range(1,N):
        r1 = (i-center1)**2+(j-center2)**2
        # euclidean distance from
        # origin is computed
        r = math.sqrt(r1)
        # using cut-off radius to
        # eliminate low frequency
        if 0 < r < d_0:
            H[i,j] = 1 - math.exp(-r**2/t1**2)

# converting H to an image
H = scipy.misc .toimage(H)
# performing the convolution
con = d * H
# computing the magnitude of the inverse FFT
e = abs(fftim.ifft2(con))
# e is converted from an ndarray to an image
f = scipy.misc .toimage(e)
# f.show()
# saving the image as ghighpass_output.png in
# Figures folder
f.save('../Figures/ghighpass_output.png')
```

The image in Figure 6.3(a) is the endothelium cell. The images in

Figures 6.3(b), 6.3(c) and 6.3(d) are the outputs of IHPF, BHPF and GHPF with cut-off radius at 30. Highpass filters are used to determine edges. Notice how the edges are formed in each case.

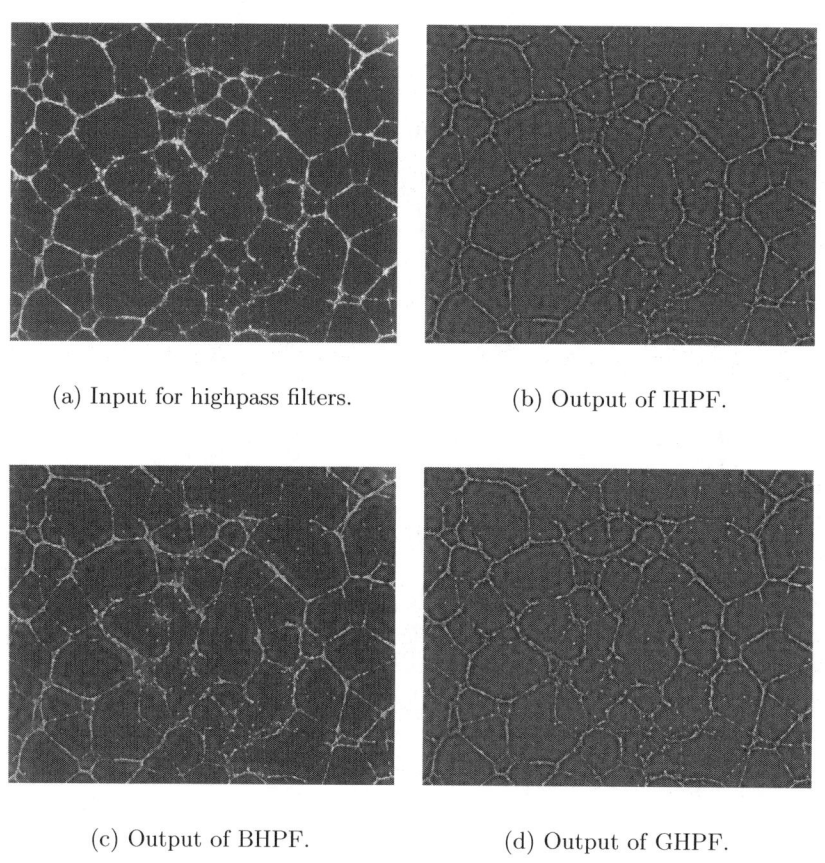

(a) Input for highpass filters.   (b) Output of IHPF.

(c) Output of BHPF.   (d) Output of GHPF.

FIGURE 6.3: An example of highpass filters. The input image and all the output images are displayed in spatial domain.

6.5.7 Bandpass Filter

A bandpass filter, as the name indicates, allows frequency from a band or range of values. All the frequencies from outside the band are set to zero. Similar to the lowpass and highpass filters, bandpass filter can be Ideal, Butterworth or Gaussian. Let us consider the ideal bandpass filter, IBPF.

The Python code for IBPF is given below.

```
import scipy.misc
import numpy, math
import scipy.fftpack as fftim
from scipy.misc.pilutil import Image

# opening the image and converting it to grayscale
a = Image.open('../Figures/fft1.png').convert('L')
# a is converted to an ndarray
b = scipy.misc.fromimage(a)
# performing FFT
c = fftim.fft2(b)
# shifting the Fourier frequency image
d = fftim.fftshift(c)

# intializing variables for convolution function
M = d.shape[0]
N = d.shape[1]
# H is defined and
# values in H are initialized to 1
H = numpy.zeros((M,N))
center1 = M/2
center2 = N/2
d_0 = 30.0 # minimum cut-off radius
d_1 = 50.0 # maximum cut-off radius

# defining the convolution function for bandpass
for i in range(1,M):
    for j in range(1,N):
        r1 = (i-center1)**2+(j-center2)**2
```

```
        # euclidean distance from
        # origin is computed
        r = math.sqrt(r1)
        # using min and max cut-off to create
# the band or annulus
        if r > d_0 and r < d_1:
            H[i,j] = 1.0


# converting H to an image
H = scipy.misc.toimage(H)
# performing the convolution
con = d * H
# computing the magnitude of the inverse FFT
e = abs(fftim.ifft2(con))
# e is converted from an ndarray to an image
f = scipy.misc.toimage(e)
# f.show()
# saving the image as ibandpass_output.png in
# Figures folder
f.save('../Figures/ibandpass_output.png')
```

The difference between this program compared to highpass or lowpass filters is in creation of the filter. In the bandpass filter, the minimum cut-off radius is set to 30 and the maximum cut-off radius is set to 50. Only intensities between 30 and 50 are passed and everything else is set to zero. Figure 6.4(a) is the input image and Figure 6.4(b) is the output image for the IBPF. Notice that the edges in the output image of IBPF is sharp compared to the input. Similar filters can be created for Butterworth and Gaussian filters using the formula discussed earlier.

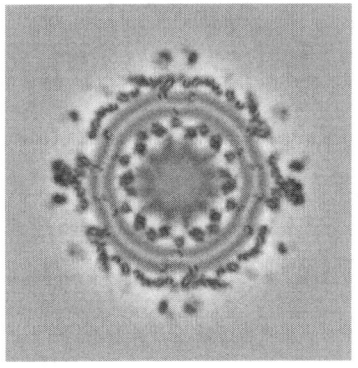 

(a) Input of IBPF.   (b) Output of IBPF.

FIGURE 6.4: An example of IBPF. The input and the output are displayed in spatial domain.

6.6 Summary

- Lowpass filters are used for noise reduction or smoothing. Highpass filters are used for edge enhancement or sharpening.

- In lowpass and highpass filters ideal, Butterworth and Gaussian were considered.

- A bandpass filter has minimum cut-off and maximum cut-off radii.

- Convolution can be viewed as the process of combining two images. Convolution is multiplication in Fourier domain. The inverse process is called deconvolution.

- Fourier transform can be used for image filtering, compression, enhancement, restoration and analysis.

6.7 Exercises

1. Fourier transform is one method for converting any function as a sum of basis functions. Perform research and find at least two other such methods. Write a report on their use in image processing.

    Hint: Wavelet, z-transform

2. An example for determining Fourier coefficient was shown earlier. However the discussion was limited to 4 coefficients. Determine the 5th coefficient assuming $f(4) = 2$.

3. The central pixel in the Fourier image is brighter compared to other pixels. Why?

4. The image in Figure 6.2(b) has a fuzzy structure next to the object. What is this called? What causes the artifact? Why are there fewer artifacts in BLPF and GLPF output images.

5. Consider an image of size 10,000-by-10,000 pixels that needs to be convolved with a filter of size 100-by-100. Comment about the most efficient method for convolving. Would it be convolution in spatial domain or Fourier?

Chapter 7

Segmentation

7.1 Introduction

Segmentation is the process of separating an image into multiple logical segments. The segments can be defined as pixels sharing similar characteristics such as intensity, texture etc. There are many methods of segmentation. They can be classified as:

- Histogram based segmentation
- Region based segmentation
- Edge segmentation
- Differential equation based method
- Model based segmentation

In this chapter, we discuss histogram and region based segmentation methods. Edge based segmentation was discussed in Chapter 4, Spatial Filters. The other two methods are beyond the scope of this book. Interested readers can refer to [28],[83] and [100] for more details.

7.2 Histogram Based Segmentation

In the histogram based method (Figure 7.1), a threshold is determined by using the histogram of the image. Each pixel in the image is compared with the threshold value. If the pixel intensity is less than the threshold value, then the corresponding pixel in the segmented image is assigned a value of zero. If the the pixel intensity is greater than the threshold value, then the corresponding pixel in the segmented image is assigned a value of 1. Thus,

if $pv \geq threshold$ **then**
　　$segpv = 1$
else
　　$segpv = 0$
end if

where pv is the pixel value in the image, $segpv$ is the pixel value in the segmented image.

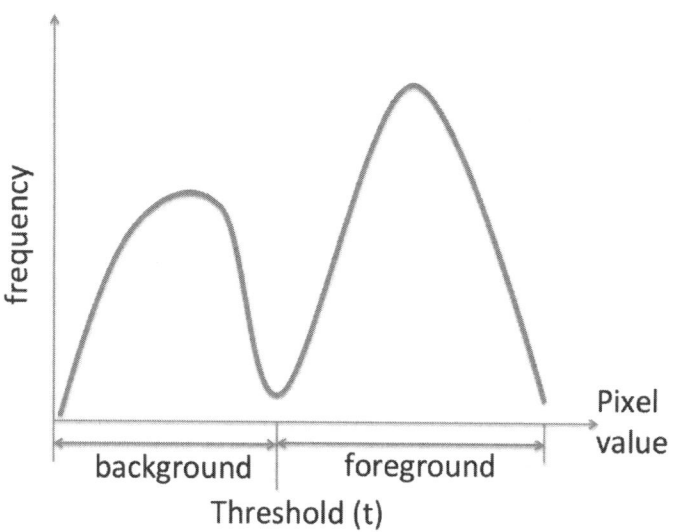

FIGURE 7.1: Threshold divides the pixels into foreground and background.

The various methods differ in their techniques of determining the threshold. There are several methods to compute the global threshold. We will discuss Otsu's method and the Renyi entropy method. In images with a non-uniform background, a global threshold value from the histogram based method might not be optimal. In such cases, local adaptive thresholding may be used.

7.2.1 Otsu's Method

Otsu's method, [72] works best if the histogram of the image is bi-modal but can be applied to other histograms as well. A bi-modal histogram is a type of histogram (similar to Figure 7.1) containing two distinct peaks separated by a valley. One peak is the background and the other foreground. Otsu's algorithm searches for a threshold value that maximizes the variance between the two groups foreground and background, so that the threshold value can better segment the foreground from the background.

Let L be the number of intensities in the image. For an 8-bit image, $L = 2^8 = 256$. For a threshold value, t the probabilities, p_i of each intensity is calculated. Then the probability of the background pixels is given by $P_b(t) = \sum_{i=0}^{t} p_i$ and the probability of foreground pixels is given by $P_f(t) = \sum_{i=t+1}^{L-1} p_i$. Let $m_b = \sum_{i=0}^{t} ip_i$, $m_f = \sum_{i=t+1}^{L-1} ip_i$ and $m = \sum_{i=0}^{L-1} ip_i$ represent the average intensities of the background, the foreground and the whole image respectively. If v_b, v_f and v represent the variance of the background, foreground and the whole image respectively. Then the variance within the groups is given by Equation 7.1 and the variance in between the groups is given by Equation 7.2.

$$v_{within} = P_b(t)v_b + P_f(t)v_f \quad (7.1)$$

$$v_{inbetween} = v - v_{within} = P_b P_f (m_b - m_f)^2. \quad (7.2)$$

For different threshold values this process of finding variance within the groups and variance between the groups is repeated. The threshold value that maximizes the variance between the groups or minimizes the variance within the group is considered the Otsu's threshold. All pixel values with intensities less than the threshold value are assigned a value of zero and all pixel values with intensities greater than the threshold value are assigned a value of one.

In the case of a color image, since there are three channels, Red, Green and Blue channel, a different threshold value for each channel is calculated.

The following is the Python function for Otsu's method:

```
skimage.filter.threshold_otsu(image, nbins=256)

Description of function arguments:

necessary argument:
image = input image in gray-scale

optional argument:
nbins = number of bins that should be considered
to calculate the histogram.
```

The Python code for Otsu's method is given below.

```
from skimage.filter.thresholding import threshold_otsu
import scipy.misc
import Image


# opening the image and converting it to grayscale
a = Image.open('../Figures/sem3.png').convert('L')
# a is converted to an ndarray
a = scipy.misc.fromimage(a)
# performing Otsu's thresholding
```

```
thresh = threshold_otsu(a)
# pixels with intensity greater than
# theshold are kept
b = a > thresh
# b is converted from ndimage to
b = scipy.misc.toimage(b)
# saving the image as sk_otsu.png
b.save('../Figures/otsu_semoutput.png')
```

In Figure 7.2(a) is a scattered electron image of an atomic element in two different phases. We segment the image using Otsu's method. The output is given in Figure 7.2(b).

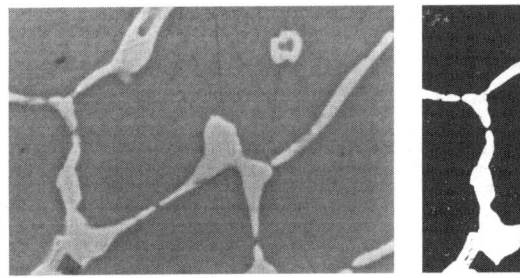
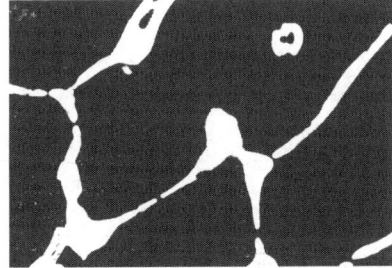

(a) Input image.   (b) Output image.

FIGURE 7.2: An example of Otsu's method. Original image reprinted with permission from Karthik Bharathwaj.

Otsu's method uses a histogram to determine the threshold and hence is very much dependent on the image. Figure 7.3(a) is an image of a spinwheel. Otsu's method is used to segment this image, and the segmented output image is shown in 7.3(b). Due to shadow on the wheel in the input image, Otsu's method did not segment the spinwheel accurately. For more on thresholding refer to [75], [84] and [94].

(a) Input image for Otsu's method.   (b) Output of Otsu's method.

FIGURE 7.3: Another example of Otsu's method.

7.2.2 Renyi Entropy

Renyi entropy based segmentation is very useful when the object of interest is small compared to the whole image i.e., the threshold is at the right tail of the histogram. For example, in the CT image of an abdomen shown in Figure 7.4(b), the tissue and background occupy more area in comparison to the bone. In the histogram, the background and tissue pixels have low pixel intensity and high frequency while the bone region has high intensity but low frequency.

In information theory and image processing, entropy quantifies the uncertainty or radomness of a variable. This concept was first introduced by Claude E. Shannon in his 1948 paper "A Mathematical Theory of Communication" [95]. This paper launched Shannon as the father of information theory. In information theory and image processing, entropy is measured in bits where each pixel value is considered as an independent random variable.

Shannon entropy is given by

$$H_1(x) = -\sum_{i=1}^{n} p(x_i) \log_a(p(x_i)) \qquad (7.3)$$

where x_i is the random variable with $i = 1, 2, ..., n$ and $p(x_i)$ is the probability of the random variable x_i and the base a can be 2, e or 10.

Alfred Renyi, a Hungarian mathematician, introduced and defined

Renyi entropy in his paper [79] in 1961. Renyi entropy is a generalization of Shannon entropy and many other entropies and is given by the following equation:

$$H_\alpha(x) = \frac{1}{1-\alpha} \log_a \left(\sum_{i=1}^{n} (p(x_i))^\alpha \right) \qquad (7.4)$$

where x_i is the random variable with $i = 1, 2, ..., n$ and $p(x_i)$ is the probability of the random variable x_i and the base a can be 2, e or 10. Renyi entropy equals Shannon entropy for $\alpha \to 1$.

The histogram of the image is used as an independent random variable to determine the threshold. The histogram is normalized by dividing each frequency with the total number of pixels in the image. This will ensure that the sum of the frequencies after normalization is one. This is the probability distribution function (pdf) of the histogram. The Renyi entropy can then be calculated for this pdf.

The Renyi entropy is calculated for all pixels below and above the threshold. These will be referred to as background entropy and foreground entropy respectively. This process is repeated for all the pixel values in the pdf. The total entropy is calculated as the sum of background entropy and foreground entropy for each pixel value in the pdf. The graph of the total entropy has one absolute maximum. The threshold value corresponding to that absolute maximum is the threshold (t) for segmentation.

The following is the Python code for Renyi entropy for an 8-bit (grayscale) image. The program execution begins with opening the CT image. The image is then processed by the function *renyi_seg_fn*. The function obtains the histogram of the image and calculates the pdf by dividing each histogram value by the total number of pixels. Two arrays, h1 and h2, are created to store the background and foreground Renyi entropy. For various thresholds, the background and foreground Renyi entropy are calculated using Equation 7.4. The total entropy is the sum of the background and foreground Renyi entropy. The threshold value for which the entropy is maximum is the Renyi entropy threshold.

```python
from scipy.misc import pilutil, fromimage
import Image
import numpy as np
from skimage.filter.thresholding
    import threshold_otsu
import skimage.exposure as imexp

# Defining function
def renyi_seg_fn(im,alpha):
    hist = imexp.histogram(im)
    # Convert all values to float
    hist_float = [float(i) for i in hist[0]]
    # compute the pdf
    pdf  = hist_float/numpy.sum(hist_float)
    # compute the cdf
    cumsum_pdf = numpy.cumsum(pdf)

    s = 0
    e = 255 # assuming 8 bit image
    scalar = 1.0/(1-alpha)
    # A very small value to prevent
    # division by zero
    eps = numpy.spacing(1)

    rr = e-s
    # The second parentheses is needed because
    # the parameters are tuple
    h1 = np.zeros((rr,1))
    h2 = np.zeros((rr,1))
    # the following loop computes h1 and h2
    # values used to compute the entropy
    for ii in range(1,rr):
        iidash = ii+s
```

```python
        temp1 = np.power(pdf[1:iidash]
                /cumsum_pdf[iidash],scalar)
        h1[ii] = np.log(numpy.sum(temp1)+eps)
        temp2 = np.power(pdf[iidash+1:255]
                /(1-cumsum_pdf[iidash]),
                scalar)
        h2[ii] = np.log(numpy.sum(temp2)+eps)

    T = h1+h2
    # Entropy value is calculated
    T = -T*scalar
    # location where the maximum entropy
    # occurs is the threshold for the renyi entropy
    location = T.argmax(axis=0)
    # location value is used as the threshold
    thresh = location
    return thresh



# Main program
# opening the image and converting it to grayscale
a = Image.open('CT.png').convert('L')
# a is converted to an ndarray
a = fromimage(a)
# computing the threshold by calling the function
thresh = renyi_seg_fn(a,3)
b = a > thresh
# b is converted from an ndarray to an image
b = pilutil.toimage(b)
# saving the image as renyi_output.png
b.save('figures/renyi_output.png')
```

Figure 7.4(a) is a CT image of abdomen. The histogram of this image is given in Figure 7.4(b). Notice that the bone region (higher pixel intensity) is on the right side of the histogram and fewer in number compared to the whole image. Renyi entropy is performed on this image to segment the bone region alone. The segmented output image is given in Figure 7.4(c).

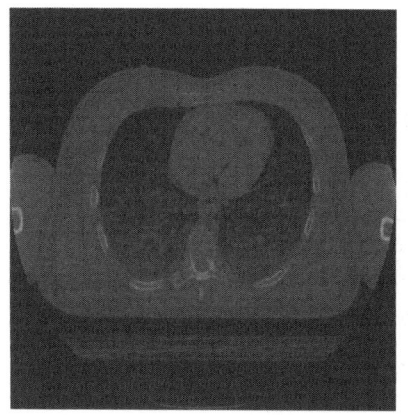

(a) Input image.

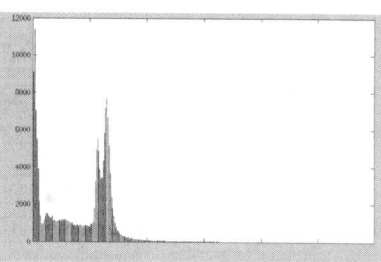

(b) Histogram of the input.

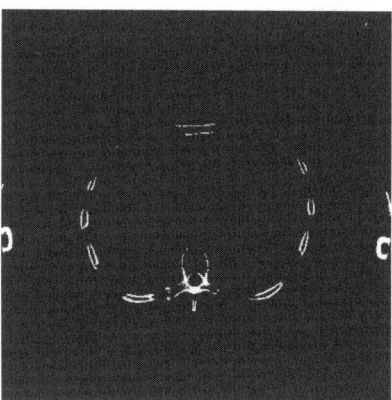

(c) Output image.

FIGURE 7.4: An example of Renyi entropy.

7.2.3 Adaptive Thresholding

As we have seen in Section 7.2.1, Otsu's Method, a global threshold might not provide accurate segmentation. Adaptive thresholding helps solve this problem. In the adaptive thresholding, the image is first divided into small sub-images. Threshold value for each sub-image is computed and is used to segment the sub-image. The threshold value for the sub-image can be computed using mean or median or Gaussian methods. In the case of mean method, the mean of sub-image is used as a threshold, while for median method, the median of the sub-image is used as a threshold. A custom formula can also be used to compute the threshold, for example we can use an average of maximum and minimum pixel values in the sub-image. By appropriate programming, any of the histogram based segmentation methods can be converted into an adaptive thresholding method.

The following is the Python function for adaptive thresholding:

```
skimage.filter.threshold_adaptive(image, block_size,
method='gaussian', offset=0, mode='reflect', param=None)

Necessary arguments:
  image is a gray-scale image

  blocksize is the size of the adaptive thresholding
  window is an integer. If it is 2, then the window size
  will be 2 by 2.

  method can be generic, gaussian, mean or median.
  For example, if method is mean, then the mean of the
  block size is used as the threshold for the window.

Optional arguments:
  offset is a floating value that should be subtracted from
```

the local threshold.

mode determines the method to handle the array border. Different options are constant, reflect, nearest, mirror, wrap.

param can be an integer or a function
If mode is guassian then the param will
be sigma (the standard deviation of the guassian).
If mode is generic, then param is a Python function.

Returns: output is a thresholded image as an ndarray.

The Python code for adaptive thresholding is given below.

```
# from skimage import filter
from skimage import filter
import scipy.misc
import Image, numpy


# opening the image and converting it to grayscale
a = Image.open('../Figures/adaptive_example1.png').
    convert('L')
# a is converted to an ndarray
a = scipy.misc.fromimage(a)
# performing adaptive thresholding
b = filter.threshold_adaptive(a,40,offset = 10)
# b is converted from an ndarray to an image
b = scipy.misc.toimage(b)
# saving the image as adaptive_output.png
# in the folder Figurespb
b.save('../Figures/adaptive_output.png')
```

In the above code, adaptive thresholding is performed using blocks of size 40-by-40. The image in Figure 7.5(a) is the input image. The lighting is non-uniform and it varies from dark on the left edge to bright on the right edge. Otsu's method uses a single threshold for the entire image and hence does not segment the image properly (Figure 7.5(b)). The text in the left section of the image is obscured by the dark region. The adaptive thresholding method (Figure 7.5(c)) uses local threshold and segments the image accurately.

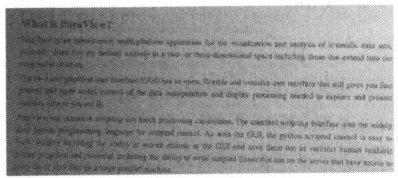
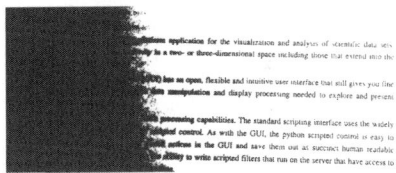

(a) Input image.   (b) Output using Otsu's method.

(c) Output using adaptive thresholding.

FIGURE 7.5: An example of thresholding with adaptive vs. Otsu's.

7.3 Region Based Segmentation

The goal of segmentation is to obtain different regions or objects in the image using methods such as thresholding. A region is a group or collection of pixels that have similar properties, so all the pixels in that region share the same characteristics. The characteristics can be pixel intensities or some other physical feature.

Previously, we have used threshold obtained from histogram to segment the image. In this section we demonstrate techniques that are based on the region of interest. In Figure 7.3, the objects are labeled as R_1, R_2, R_3, R_4 and the background as R_5.

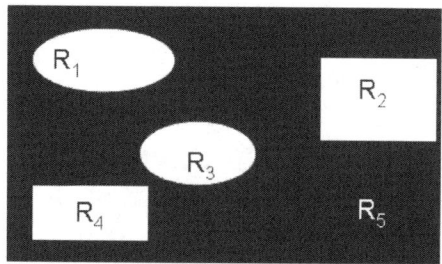

FIGURE 7.6: An example of an image for region-based segmentation.

The different regions constitute the image, $\bigcup_{i=1}^{5} R_i = I$ where I represents the whole image. No two regions overlap, $R_i \cap R_j = \emptyset$ for $i \neq j$. Every region is connected, with I representing the image and R_i representing the regions for $i = 1$ to n. We can now formulate basic rules that govern the region-based segmentation.

1. All the regions combined should equal the image, $\bigcup_{i=1}^{n} R_i = I$

2. Each region, R_i is connected for $i = 1$ to n

3. No two regions overlap, $R_i \cap R_j = \emptyset$

To segment the regions, we need some a priori information. This a priori information is the seed pixels, pixels that are part of the foreground. The seed pixels grow by considering the pixels in their neighborhood that have similar properties. This process connects all the pixels in a region with similar properties. The region growing process will terminate when there are no more pixels to add that share the same characteristics of the region.

It might not always be possible to have a priori knowledge of the seed pixels. In such cases, a list of characteristics of different regions should be considered. Then pixels that satisfy the characteristics of a particular region will be grouped together. The most popular region-based segmentation method is the watershed segmentation.

7.3.1 Watershed Segmentation

To perform watershed segmentation, a grayscale image is considered. The grayscale values of the image represent the peaks and valleys of the topographic terrain of the image. The lowest valley in an object is the absolute minumum. The highest grayscale value corresponds to the highest point in the terrain. The watershed segmentation can be explained as follows: all the points in a region where if a drop of water was placed will settle to the absolute minimum are known as the catchment basin of that minimum or watershed. If water is supplied at a uniform rate from the absolute minimum in an object, as water fills up the object, at some point water will overflow into other objects. Dams are constructed to stop water from overflowing into other objects/regions. These dams are the watershed segmentation lines. The watershed segmentation lines are edges that separate one object from another. More details on watershed lines can be found in [64].

Now let us look at how the dams are constructed. For simplicity, let us assume that there are two regions. Let R_1 and R_2 be two regions and let C_1 and C_2 be the corresponding catchment basins. Now for each time step, the regions that constitute the catchment basins are increased. This can be achieved by dilating the regions with a structuring element of size 3-by-3 (say). If C_1 and C_2 become one connected region in the time step n, then at the time step $n-1$ the regions C_1 and C_2 were disconnected. The dams or the watershed lines can be obtained by taking the difference of images at time steps n and $n-1$.

In 1992, F. Meyer proposed an algorithm to segment color images, [63]. Internally, cv2.waterhsed uses Meyer's flooding algorithm to perfrom watershed segmentation. Meyer's algorithm is outlined below:

1. The original input image and the marker image are given as inputs.

2. For each region in the marker image, its neighboring pixels are placed in a ranked list according to their gray levels.

3. The pixel with the highest rank (highest gray level) is compared with the labeled region. If the pixels in the labeled region have same gray level as the given pixel, then the pixel is included in the labeled region. Then a new ranked list with the neighbors is formed. This step contributes towards the growing of the labeled region.

4. The above step is repeated until there are no elements in the list.

Prior to performing watershed, the image has to be preprocessed to obtain a marker image. Since the water is supplied from catchment basins, these basin points are guaranteed foreground pixels. The guaranteed foreground pixel image is known as the marker image.

The preprocessing operations that should be performed before watershed are as follows:

1. Foreground pixels are segmented from the background pixels.

2. Erosion is performed to obtain foreground pixels only. Erosion is a morphological operation in which the background pixels grow and foreground pixels shrink. Erosion is explained in detail in Chapter 8, Morphological Operations in Sections 8.4 and 8.5.

3. Distance transform creates an image where every pixel contains the value of the distance between itself and the nearest background pixel. Thresholding is done to obtain the pixels that are

farthest away from the background pixels and are guaranteed to be foreground pixels.

4. All the connected pixels in a region are given a value in the process known as labeling. The labeled image is used as a marker image. Further explanation on labeling can be found in Chapter 9, Image Measurements in Section 9.2.

These operations along with the watershed are used in the cv2.watershed code provided below.

All the cv2 functions that are used for preprocessing such as erode, threshold, distance transform, and watershed are explained below. A more detailed documentation can be found at [69]. This will be followed by the Python program using cv2 module.

The cv2 function for erosion is as follows:

```
cv2.erode(input,element,iterations,anchor,
borderType, borderValue)

Necessary arguments:

input is the input image.

iterations is an integer value corresponding to the
number of times erosion is performed.

Optional arguments:

element is the structuring element. The default value
is None.

If element is specified, then anchor is the center of
the element. The default value is (-1,-1).
```

borderType is similar to mode argument in convolve function.

If borderType is constant then borderValue should be specified.

Returns: An eroded image.

The cv2 function for thresholding is given below:

---

cv2.threshold(input,thresh,maxval,type)

Necessary arguments:

input is an input array. It can be either 8 bit or 32 bit.

thresh is the threshold value.

Optional arguments:

maxval should be assigned and will be used when the threshold type is THRESH_BINARY or THRESH_BINARY_INV.

type can be either THRESH_BINARY, THRESH_BINARY_INV, THRESH_TRUNC, THRESH_TOZERO or THRESH_TOZERO_INV.
Also, THRESH_OTSU can be added to any of the above.
For example, in THRESH_BINARY+THRESH_OTSU the threshold value is determined by Otsu's method and then that threshold value will be by THRESH_BINARY. The pixels with intensities greater than the threshold value will be assigned the maxval

and rest will be assigned 0.

Returns: Output array same size and type as input array.

The cv2 function for distance transform is given below:

---

cv2.DistTransform(image, distance_Type, mask_Size, labels, labelType)

Necessary arguments:

image is a 8-bit single channel image.

distance_Type is used to specify the distance formula. It can be either CV_DIST_L1 (given by 0), CV_DIST_L2 (given by 1) or CV_DIST_C (given by 2). The distance between (x,y) and (t,s) for CV_DIST_L1 is |x-t|+|y-s| while CV_DIST_L2 is the Euclidean distance and CV_DIST_C is the max{|x-t|,|y-s|}.

The size for the mask can be specified by mask_Size.
If mask_Size is 3, a 3-by-3 mask is considered.

Optional arguments:

A 2D array of labels can be returned using labels.

The type of the above array of labels can be specified by labelType. If labelType is DIST_LABEL_CCOMP, then each connected component will be assigned the same label. If labelType is DIST_LABEL_PIXEL then each connected component will have its own label.

Returns: Output is a distance image same size as the input.

The cv2 function for watershed is given below:

---

cv2.watershed(image, markers)

Necessary arguments:
    image is the 8-bit 3 channel color image. Internally, the function converts the color image to grayscale. Only accepts color image as input.

markers is a labelled 32-bit single channel image.

Returns:
    Output is a 32 bit image. Output is overwritten on the marker image.

The cv2 code for the watershed segmentation is given below. The various Python statements leading to the watershed function create the marker image. The image in Figure 7.7(a)) are dyed osteoblast cells cultured in a bottle. The image is read and thresholded (Figure 7.7(b)) to obtain foreground pixels. The image is converted to a grayscale image before thresholding. The image is eroded (Figure 7.7(c)) to ensure that guaranteed foreground pixels are obtained. Distance transform (Figure 7.7(d)) and the corresponding thresholding (Figure 7.7(e)) ensures the guaranteed foreground pixel image (i.e., marker image) is obtained. The marker image is used in watershed to obtain the image shown in Figure 7.7(f). The inputs for cv2 watershed function are input image as a color image and a marker image.

```
import cv2
```

```
from scipy.ndimage import label
import scipy.misc
import Image, numpy
# from skimage.morphology  import label


# opening the image and converting it to grayscale
a = cv2.imread('../Figures/cellimage.png')
# covnerting image from color to grayscale
a1 = cv2.cvtColor(a, cv2.COLOR_BGR2GRAY)
# thresholding the image to obtain cell pixels
thresh,b1 = cv2.threshold(a1, 0, 255,
            cv2.THRESH_BINARY_INV+cv2.THRESH_OTSU)
# since Otsu's method has over segmented the image
# erosion operation is performed
b2 = cv2.erode(b1, None,iterations = 2)
# distance transform is performed
dist_trans = cv2.distanceTransform(b2, 2, 3)
# thresholding the distance transform image to obtain
# pixels that are foreground
thresh, dt = cv2.threshold(dist_trans, 1,
             255, cv2.THRESH_BINARY)
# performing labeling
#labelled = label(b, background = 0)
labelled, ncc = label(dt)
# labelled is converted to 32-bit integer
labelled = labelled.astype(numpy.int32)
# performing watershed
cv2.watershed(a, labelled)
# converting the ndarray to image
dt1 = scipy.misc.toimage(labelled)
# saving the image as watershed_output.png
dt1.save('../Figures/watershed_output.png')
```

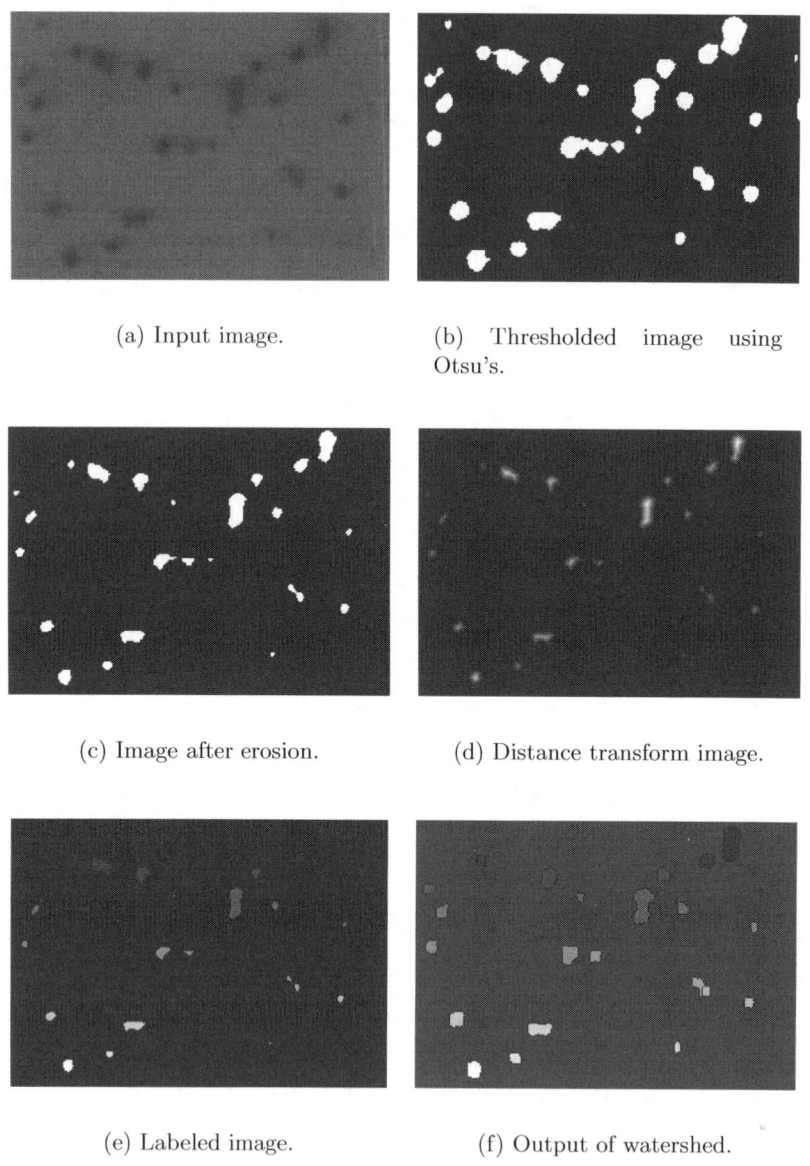

(a) Input image.

(b) Thresholded image using Otsu's.

(c) Image after erosion.

(d) Distance transform image.

(e) Labeled image.

(f) Output of watershed.

FIGURE 7.7: An example of watershed segmentation. Original image reprinted with permission from Dr. Susanta Hui, Masonic Cancer Center, University of Minnesota.

7.4 Segmentation Algorithm for Various Modalities

So far we have discussed a few segmentation algorithms without concerning ourselves with the imaging modalities. Each imaging modality has unique characteristics that need to be understood in order to create a good segmentation algorithm.

7.4.1 Segmentation of Computed Tomography Image

The details of CT imaging are discussed in Chapter 10, Computer Tomography. In a CT image, the pixel intensities are in Hounsfield unit. The pixel intensities have physical significance as they are a map of the electron density of that material. The units are the same whether we image a human being, a mouse or a dog. Thus, a pixel intensity of $+1000$ always corresponds to a material that has electron density similar to bone. A pixel intensity of -1000 always corresponds to a material that has electron density similar to air. Hence segmentation process becomes simpler in the case of CT. To segment bone in a CT image, a simple thresholding such as assigning all pixels with values greater than $+1000$ being assigned 1 will suffice. A list of the range of pixel values corresponding to various materials such as soft tissue, hard tissue etc. have been created and hence simplify the segmentation process. This however assumes that the CT image has been calibrated to a Hounsfield unit. If not, traditional segmentation techniques have to be used.

7.4.2 Segmentation of MRI Image

The details of MRI are discussed in Chapter 11, Magentic Resonance Imaging. MRI images do not have a standardized unit and hence need to be segmented using more traditional segmentation techniques.

7.4.3 Segmentation of Optical and Electron Microscope Image

The details of optical and electron microscope are discussed in Chapter 12, Optical Microscope and Chapter 13, Electron Microscope, respectively. In CT and MRI imaging of patients, the shape, size and position of organs remain similar across patients. In the case of optical and electron microscope, two images acquired from the same specimen may not look alike and hence traditional techniques have to be used.

7.5 Summary

- Segmentation is a process of separating an image into multiple logical segments.

- Histogram based and region based segmentation methods were discussed.

- Histogram based method determines the threshold based on histogram.

- Otsu's method determines the threshold that maximizes the variance between the groups or minimizes the variance within the group.

- The threshold that maximizes the entropy between the foreground and background is the Renyi entropy threshold.

- Adaptive thresholding method segments the image by dividing the image into sub-images and then applying thresholding to each sub-image.

- Watershed segmentation is used when there are overlapping objects in an image.

7.6 Exercises

1. In this chapter, we discussed a few segmentation methods. Consult the books listed as reference and explain at least three more methods including details of the segmentation process, its advantages and disadvantages.

2. Consider any of the images used in histogram based segmentation in this chapter. Rotate or translate the image using ImageJ by various angles and distance, and for each case segment the image. Are the threshold values different for different levels of rotation and translation? If there are differences in threshold value, explain the cause of the changes.

3. What happens if you zoom into the image using ImageJ while keeping the image size the same? Try different zoom levels (2X, 3X, and 4X). Explain the cause of change in threshold value.

   Hint: This changes the content of the image significantly and hence the histogram and the segmentation threshold.

4. In the various segmentation results, you will find spurious objects. Suggest a method to remove these objects.

   Hint: Morphology.

Chapter 8

Morphological Operations

8.1 Introduction

So far, we have discussed the various methods for manipulating individual pixels in the image through filtering, Fourier transform etc. An important part of image analysis involves understanding the shape of the objects in that image through morphological operations. Morphology means form or structure. In morphological operations, the goal is to transform the structure or form of the objects using a structuring element. These operations change the shape and size of the objects in the image. Morphological operations can be applied on binary, grayscale and color images. We omit color morphology in this chapter, as most bio-medical images are grayscale or binary images. We begin with basic morphological operations such as dilation, erosion, opening, and closing and then progress to compound operations such as hit-or-miss and skeletonization.

8.2 History

Morphology was introduced by Jean Serra in the 1960s as a part of his Ph.D. thesis under Georges Matheron at the Ecole des Mines de Paris, France. Serra applied the techniques he developed in the field of geology. With the arrival of modern computers, morphology began to be applied on images of all types such as black and white, grayscale and color. Over the next several decades, Serra developed the formalism for applying morphology on various data types like images, videos, meshes etc. More information can be found in [19],[34],[65],[67],[92],[93],[98].

8.3 Dilation

Dilation allows the foreground pixels in an image to grow or expand. This operation will also fill small holes in an object. It is also used to combine objects that are close enough to each other but are not connected.

The dilation of the image I with a structuring element S is denoted as $I \oplus S$.

Figure 8.1(a) is a binary image of size 4-by-5. The foreground pixels have intensity of 1 while background pixels have intensity of 0. The structuring element, Figure 8.1(b), is used to perform the dilation. The dilation process is explained in detail in the following steps:

1. Figure 8.1(a) is the binary image with 0's and 1's as the input.

2. The structuring element that will be used for dilation is shown in Figure 8.1(b). The X on the 1 represents the reference pixel or origin of the structuring element. In this case the structuring element is of size 1-by-2. Both values in the structuring element play an important role in the dilation process.

3. To better illustrate the dilation process, we consider the first row in Figure 8.1(a) and apply the structuring element on each pixel in that row.

4. With this structuring element, we can only grow the boundary by one more pixel to the right. If we considered a 1-by-3 structuring element with all 1's and the origin of the structuring element at

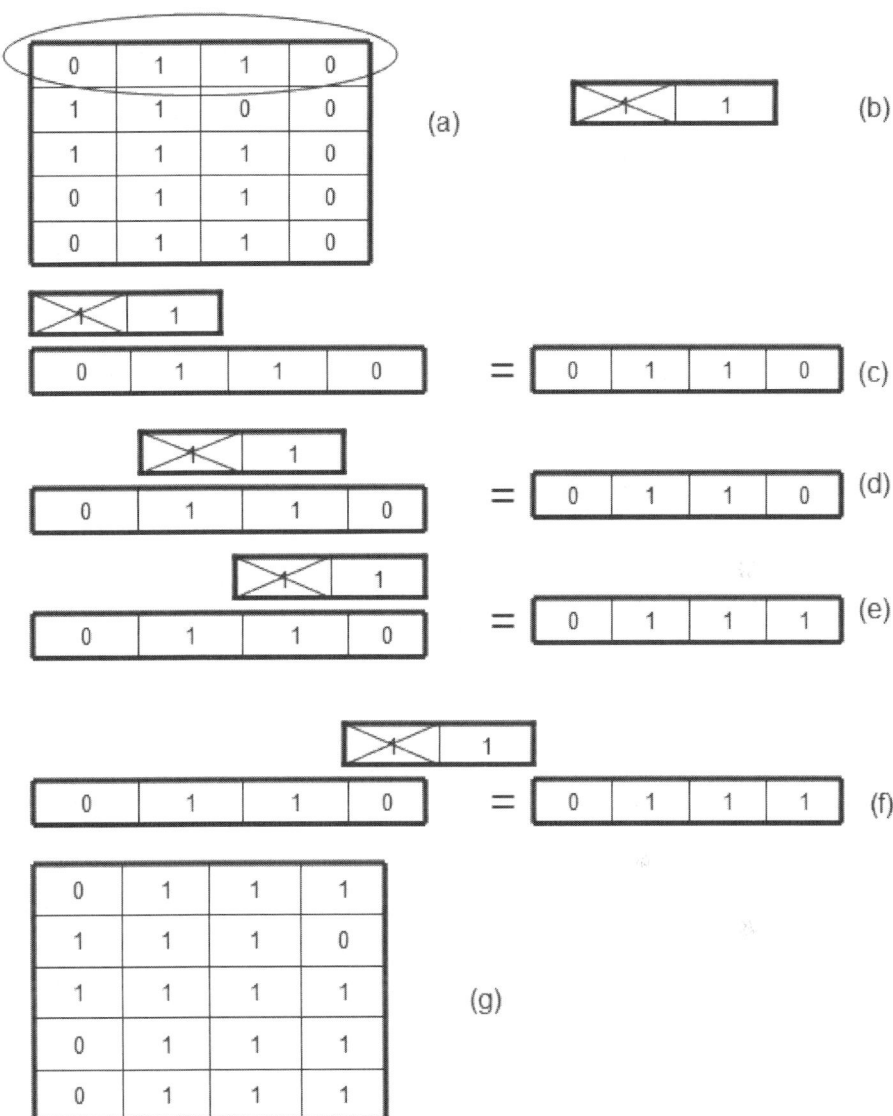

FIGURE 8.1: An example of binary dilation: (a) binary image for dilation, (b) structuring element, (c), (d), (e), (f) application of dilation at various points and (g) final output after dilation.

the center, then the boundary will grow by one pixel each in the left and right directions. Note that the morphological operations are performed on the input image and not on the intermediate results. The output of the morphological operation is the aggregate of all the intermediate results.

5. The structuring element is placed over the first pixel of the row and the pixel values in the structuring element are compared with the pixel values in the image. Since the reference value in the structuring element is 1 whereas the underlying pixel value in the image is 0, the pixel value in the output image remains unchanged. In Figure 8.1(c) the left side is the input to the dilation process and the right side is the intermediate result.

6. The structuring element is then moved one pixel over. Now the reference pixel in the structuring element and the image pixel value match. Since the next value to the reference value also matches with the 1 in the underlying pixel value, the pixel values in the output image do not change, and the output is shown in Figure 8.1(d).

7. The structuring element is then moved one pixel over. Now the reference pixel in the structuring element and the pixel value match. But the next value to the reference value does not match with the 0 in the underlying pixel value, the pixel value in the intermediate result will be changed to 1 as in Figure 8.1(e).

8. Then the structuring element is then moved one pixel over. Since the structuring element is out of image bound, there is no change in the underlying image pixel value as shown in Figure 8.1(f).

9. This process is repeated on every pixel in the input image. The output of the dilation process on the whole image is given in Figure 8.1(g).

10. The process can be iterated multiple times using the same structuring element. In such case, the output from the previous iteration (Figure 8.1(g)) is used as input to the next iteration.

In summary, the dilation process first detects the boundary pixels of the object and it grows the boundary by certain number of pixels (1 pixel to the right in this case). By repeating this process through multiple iterations or by using a large structuring element, the boundary pixels can grow by several pixels.

The following is the Python function for binary dilation:

```
scipy.ndimage.morphology.binary_dilation(input,
    structure=None,iterations=1,mask=None,
    output=None,border_value=0,
    origin=0,brute_force=False)
```

Necessary arguments:
  input = input image

Optional arguments:
  structure is the structuring element used for the dilation, which was discussed earlier. If no structure is provided, scipy assumes a square structuring element of value 1.
  The data type is ndarray.

  iterations are the number of times the dilation operation is repeated. The default value is 1. If the value is less than 1, the process is repeated until there is no change in results.
  The data type is integer or float.

  mask is an image, with the same size as the input image with value of either 1 or 0. Only points in the input image corresponding to value of 1 in the mask image are modified at each iteration. This is useful, if only a portion of the input image needs to be dilated. The data type is an ndarray.

origin determines origin of the structuring element, structure. The default value 0 corresponds to a structuring element whose origin (reference pixel) is at the center. The data needs to be either int for 1D structuring element or tuples of int for multiple dimension. Each value in the tuple corresponds to different dimensions in the structuring element.

border_value will be used for the border pixels in the output image. It can either be 0 or 1.

Returns: output as an ndarray.

The following is Python code that takes an input image and performs dilation with 5 iterations using the binary_dilation function:

```
from scipy.misc import toimage
import Image
import scipy.ndimage as snd

# opening the image and converting it to grayscale
a = Image.open('../figures/dil_image.png').
    convert('L')
# performing binary dilation for 5 iterations
b = snd.morphology.binary_dilation(a,iterations=5)
# converting b from an ndarray to an image
b = toimage(b)
# displaying the image
```

```
b.show()
```

Figure 8.2(a) is the input image for binary dilation with 5 iterations and the corresponding output image is given in Figure 8.2(b). Since binary dilation makes the foreground pixels dilate or grow, the small black spots (background pixels) inside the white regions (foreground pixels) in the input image disappear.

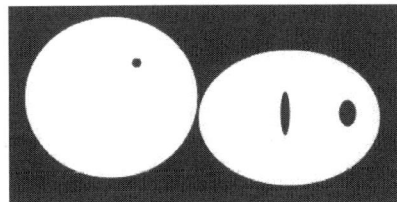 

(a) Black and white image for dilation.

(b) Output image after dilation with 5 iterations.

FIGURE 8.2: An example of binary dilation.

8.4 Erosion

Erosion is used to shrink objects in an image by removing pixels from the boundary of that object. Erosion is opposite of dilation.

The erosion of the image I and with a structuring element S is denoted as $I \ominus S$.

Let us consider the same binary input and the structuring element that was considered for dilation to illustrate erosion. Figure 8.3(a) is a binary image of size 4 by 5. The structuring element 8.3(b) is used to perform the erosion. The erosion process is explained in detail in the following steps:

1. Figure 8.3(a) is an example of a binary image with 0's and 1's.

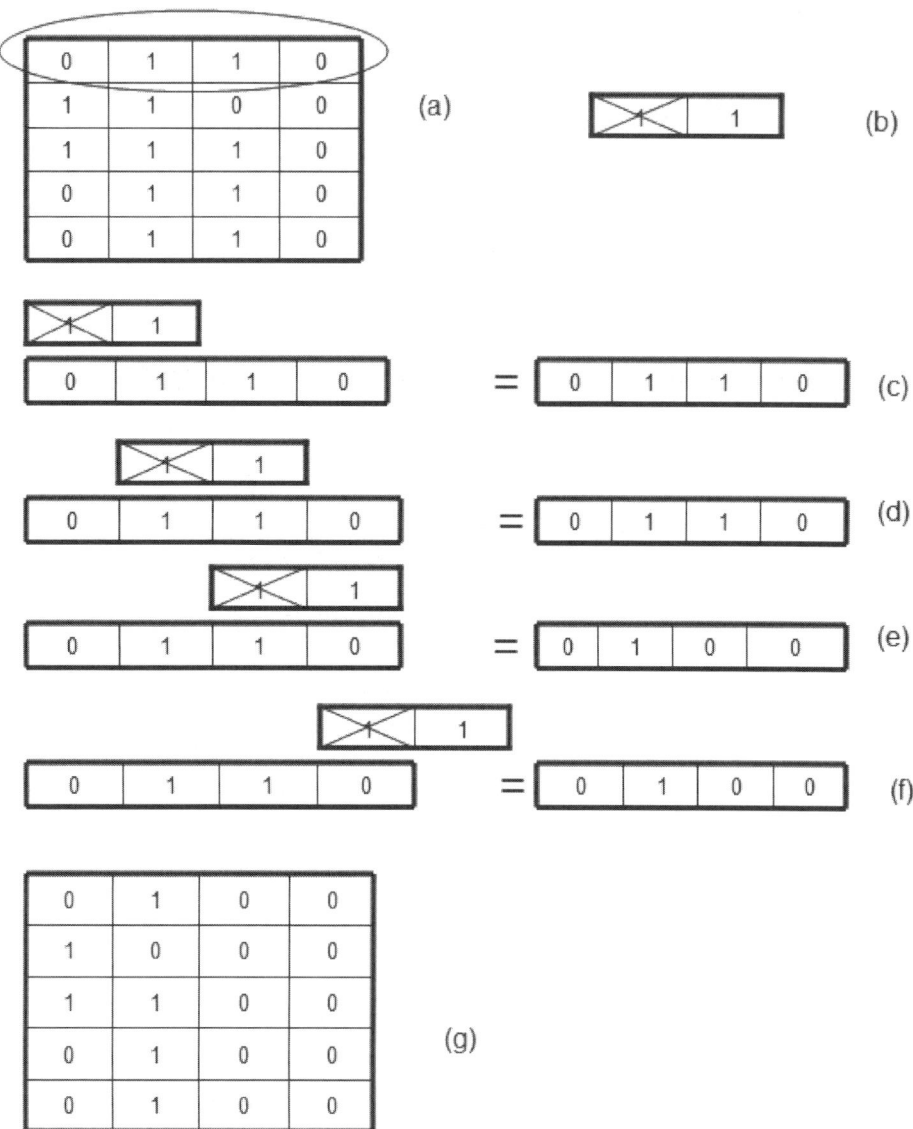

FIGURE 8.3: An example of binary erosion: (a) binary image for erosion, (b) structuring element, (c), (d), (e), (f) application of erosion at various points and (g) final output after erosion.

The background pixels are represented by 0 and the foreground by 1.

2. The structuring element that will be used for erosion is shown in Figure 8.3(b). The X on the 1 represents the reference pixel in the structuring element. In this case the structuring element is of size 1-by-2. Both the values in the structuring element play an important role in erosion.

3. Consider the first row in Figure 8.3(a) and apply the structuring element on each pixel of the row.

4. With this structuring element, we can only erode the boundary by one pixel to the right.

5. The structuring element is placed over the first pixel of that row and the pixel values in the structuring element are compared with the pixel values in the image. Since the reference value in the structuring element is 1 whereas the underlying pixel value in the image is 0, the pixel value remains unchanged. In Figure 8.3(c), the left side is the input to the erosion process and the right side is the intermediate output.

6. The structuring element is then moved one pixel over. The reference pixel in the structuring element and the image pixel value match. Since the next value to the reference value also matches with the 1 in the underlying pixel value, the pixel values in the output image do not change, as shown in Figure 8.3(d).

7. The structuring element is then moved one pixel over. The reference pixel in the structuring element and the image pixel value match but the non-reference value does not match with the 0 in the underlying pixel value. The structuring element is on the boundary. Hence, the pixel value below the reference value is replaced with 0, as shown in Figure 8.3(e).

8. The structuring element is then moved one pixel over. Since the structuring element is out of image bound, there is no change in the underlying image pixel value, as shown in Figure 8.3(f).

9. This process is repeated on every pixel in the input image. The output of the erosion process on the whole image is given in Figure 8.3(g).

10. The process can be iterated multiple times using the same structuring element. In such case, the output from the previous iteration (Figure 8.3(g)) is used as input to the next iteration.

In summary, the erosion process first detects the boundary pixels of the object and shrinks the boundary by a certain number of pixels (1 pixel from the right in this case). By repeating this process through multiple iterations or by using a larger structuring element, the boundary pixels can be shrunk by several pixels.

The Python function for binary erosion is given below. The arguments for binary erosion are the same as the binary dilation arguments listed previously.

```
scipy.ndimage.morphology.binary_erosion(input,
  structure=None,iterations=1,mask=None,
  output=None,border_value=0,origin=0,
  brute_force=False)
```

The Python code for binary erosion is given below.

```
from scipy.misc import toimage, fromimage
import Image
import scipy.ndimage as snd

# opening the image and converting it to grayscale
a = Image.open('../figures/er_image.png').
    convert('L')
# performing binary erosion for 5 iterations
b = snd.morphology.binary_erosion(a,iterations=25)
```

```
# converting b from an ndarray to an image
b = toimage(b)
# displaying the image
b.show()
```

Figures 8.4(b) and 8.4(c) demonstrate the binary erosion of 8.4(a) using 10 and 20 iterations respectively. Erosion removes boundary pixels and, hence after 10 iterations the two circles are separated creating a dumbbell shape. A more profound dumbbell shape is obtained after 20 iterations.

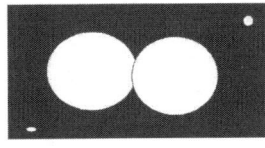

(a) Input image for erosion.

(b) Output image after 10 iterations.

(c) Output image after 20 iterations.

FIGURE 8.4: An example of binary erosion.

8.5 Grayscale Dilation and Erosion

Grayscale dilation and erosion are similar to their binary counterparts. In grayscale dilation, bright pixels increase or grow and dark pixels decrease or shrink. The effect of dilation can be clearly observed in a region(s) where there is a change in the grayscale intensity. Similar to binary dilation, grayscale dilation fills holes.

In grayscale erosion, the bright pixel values will shrink and the dark pixels increase or grow. Small bright objects will be eliminated by grayscale erosion and dark objects will grow. As in the case of dilation, the effect of erosion can be observed in region(s) where there is a change in the grayscale intensity.

8.6 Opening and Closing

Opening and closing operations are complex morphological operations. They are obtained by combining dilation and erosion. Opening and closing can be performed on binary, grayscale and color images.

Opening is defined as erosion followed by dilation of an image. The opening of the image I with a structuring element S is denoted as

$$I \circ S = (I \ominus S) \oplus S \tag{8.1}$$

Closing is defined as dilation followed by erosion of an image. The closing of the image I with a structuring element S is denoted as

$$I \bullet S = (I \oplus S) \ominus S \tag{8.2}$$

The following is the Python function for opening:

```
scipy.ndimage.morphology.binary_opening(input,
structure=None, iterations=1, output=None, origin=0)

Necessary arguments:
input = array

Optional arguments:
    structure is the structuring element used for
    the dilation, which was discussed earlier. If no
    structure is provided, scipy assumes a square
    structuring element of value 1.
```

The data type is ndarray.

iterations are the number of times the opening is performed (erosion followed by dilation). The default value is 1. If the value is less than 1, the process is repeated until there is no change in results. The data type is integer or float.

origin determines origin of the strcuturing element, structure2. The default value 0 corresponds to a structuring element whose origin (reference pixel) is at the center. The data needs to be either int for 1D structuring element or tuples of int for multiple dimension. Each value in the tuple corresponds to different dimensions in the structuring element.

Returns: output as an ndarray

The Python code for binary opening with 5 iterations is given below.

```
from scipy.misc import toimage
import Image
import scipy.ndimage as snd

# opening the image and converting it to grayscale
a = Image.open('../figures/dil_image.png').
    convert('L')
# defining the structuring element
s = [[0,1,0],[1,1,1], [0,1,0]]
# performing the binary opening for 5 iterations
b = snd.morphology.binary_opening(a, structure=s,
    iterations=5)
```

```
# b is converted from an ndarray to an image
b = toimage(b)
# displaying the image
b.show()
```

Figure 8.5(b) is the output of the binary opening with 5 iterations. Binary opening has altered the boundaries of the foreground objects. The size of the small black holes inside the objects has also changed.

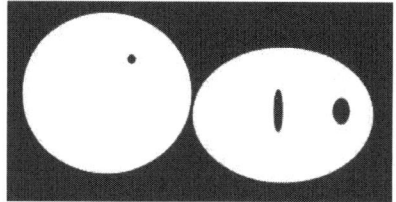
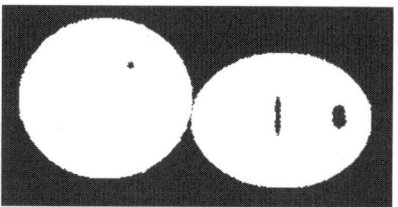

(a) Input image for opening. (b) Output image after opening.

FIGURE 8.5: An example of binary opening with 5 iterations.

The Python function for binary closing is given below. The arguments for binary closing are the same as the binary opening arguments.

```
scipy.ndimage.morphology.binary_closing(input,
   structure=None, iterations=1,output=None, origin=0)
```

The Python code for closing is given below and an example is given in Figure 8.6. The closing operation has resulted in filling in the holes, as shown in Figure 8.6(b).

```
from scipy.misc import toimage, fromimage
import Image
import scipy.ndimage as snd
```

```
# opening the image and converting it to grayscale
a = Image.open('../figures/dil_image.png').
    convert('L')
# defining the structuring element
s = [[0,1,0],[1,1,1], [0,1,0]]
# performing the binary closing for 5 iterations
b = snd.morphology.binary_closing(a,structure=s,
    iterations=5)
b = toimage(b)
b.show()
```

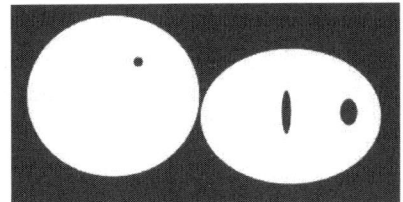 

(a) Input image for closing. (b) Output image after closing.

FIGURE 8.6: An example of binary closing with 5 iterations.

It can be observed that the black holes in the input image are elongated after the opening operation, while the closing operation on the same input filled the holes.

8.7 Hit-or-Miss

Hit-or-miss transformation is a morphological operation used in finding specific patterns in an image. Hit-or-miss is used to find boundary or corner pixels, and is also used for thinning and thickening, which are discussed in the next section. Unlike the methods we have discussed so far, this method uses more than one structuring element and all its variations to determine pixels that satisfy a specific pattern.

Let us consider a 3-by-3 structuring element with origin at the center. The structuring element with 0's and 1's shown in Table 8.1 is used in the hit-or-miss transformation to determine the corner pixels. The blank space in the structuring element can be filled with either 1 or 0.

	1	
0	1	1
0	0	

TABLE 8.1: Hit-or-miss structuring element

Since we are interested in finding the corner pixels we have to consider all the four variations of the structuring element in Table 8.1. The four structuring elements given in Table 8.2 will be used in the hit-or-miss transformation to find the corner pixels. The origin of the structuring element is applied to all pixels in the image and the underlying pixel values are compared. As discussed in Chapter 4 on filtering, the structuring element cannot be applied to the edges of the image. So the edges of the image are assumed to be zero in the output.

After determining the locations of the corner pixels from each structuring element, the final output of hit-or-miss is obtained by performing an OR operation on all the output images.

	1	
0	1	1
	0	

	1	
1	1	0
	0	0

0	0	
0	1	1
	1	

0	0	
1	1	0
	1	

TABLE 8.2: Variation of all structuring elements used to find corners.

Let us consider a binary image in Figure 8.7(a). After performing the hit-or-miss transformation on this image with the structuring elements in Table 8.2, we obtain the image in Figure 8.7(b). Notice that the pixels in the output of Figure 8.7(b) are a subset of boundary pixels.

0	0	0	0	0	0	0	0	0	0	0
0	0	0	1	0	1	0	0	0	0	0
0	0	1	1	0	1	0	0	0	0	0
0	0	1	1	1	1	1	0	0	0	0
0	0	1	1	1	1	1	0	0	0	0
0	0	1	0	0	0	1	1	0	0	0
0	0	1	0	0	0	1	1	0	0	0
0	0	1	1	1	1	1	0	0	0	0
0	0	1	1	1	1	0	0	0	0	0
0	0	0	0	0	0	0	0	0	0	0
0	0	0	0	0	0	0	0	0	0	0

(a) Input image for hit-or-miss.

0	0	0	0	0	0	0	0	0	0	0
0	0	0	0	0	0	0	0	0	0	0
0	0	1	0	0	0	0	0	0	0	0
0	0	0	0	0	0	1	0	0	0	0
0	0	0	0	0	0	1	0	0	0	0
0	0	0	0	0	0	0	1	0	0	0
0	0	0	0	0	0	0	1	0	0	0
0	0	0	0	0	0	1	0	0	0	0
0	0	1	0	0	0	0	0	0	0	0
0	0	0	0	0	0	0	0	0	0	0
0	0	0	0	0	0	0	0	0	0	0

(b) Output image of hit-or-miss.

FIGURE 8.7: An example of hit-or-miss transformation.

The following is the Python function for hit-or-miss transformation:

```
scipy.ndimage.morphology.binary_hit_or_miss(input,
  structure1=None, structure2=None,
output=None, origin1=0, origin2=None)

Necessary arguments:
input is a binary array

Optional arguments:
  structure1 is a structuring element that is used
  to fit the foreground of the image.  If no
  structuring element is provided, then scipy will
  assume square structuring element of value 1.

  structure2 is a structuring element that is used
  to miss the foreground of the image. If no
  structuring element is provided, then scipy will
  consider a complement of structuring element
  provided in structure1.
```

origin1 determines origin of the structuring element, structure1. The default value 0 corresponds to a structuring element whose origin (reference pixel) is at the center. The data needs to be either int for 1D structuring element or tuples of int for multiple dimension. Each value in the tuple corresponds to different dimensions in the structuring element.

origin2 determines origin of the structuring element, structure2. The default value 0 corresponds to a structuring element whose origin (reference pixel) is at the center. The data needs to be either int for 1D structuring element or tuples of int for multiple dimension. Each value in the tuple corresponds to different dimensions in the structuring element.

Returns: output as an ndarray.

The Python code for hit-or-miss transform is given below.

```
from scipy.misc import toimage, fromimage
import Image
import numpy as np
import scipy.ndimage as snd

# opening the image and converting it to grayscale
a = Image.open('../figures/thickening_input.png').
    convert('L')
# defining the structuring element
structure1 = np.array([[1, 1, 0], [1, 1, 1],
            [1, 1, 1]])
# performing the binary hit-or-miss
```

```
b = snd.morphology.binary_hit_or_miss(a,
    structure1=structure1)
# b is converted from an ndarray to an image
b = toimage(b)
# displaying the image
b.show()
```

In the above program, a structuring element 'structure1' is created with all the elements listed and used in the hit-or-miss transformation. Figure 8.8(a) is the input image for hit-or-miss transform and the corresponding output is in Figure 8.8(b). Notice that only few boundary pixels from each object in the input image are identified by the hit-or-miss transformation. It is important to make a judicious choice of the structuring element in the hit-or-miss transform, as different elements have different effect on the output.

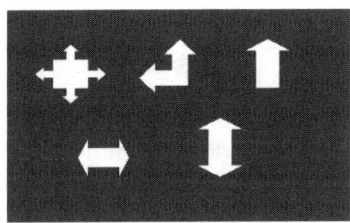
(a) Input image for hit-or-miss

(b) Output image of hit-or-miss

FIGURE 8.8: An example of hit-or-miss transformation on a binary image.

8.8 Thickening and Thinning

Thickening and thinning transformations are an extension of hit-or-miss transformation and can only be applied to binary images.

Thickening is used to grow the foreground pixels in a binary image and is similar to the dilation operation. In this operation, the background pixels are added to the foreground pixels to make the selected region grow or expand or thicken. The thickening operation can be expressed in terms of the hit-or-miss operation. Thickening of the image I with the structuring element S can be given by Equation 8.3 where H is the hit-or-miss on image I with S,

$$\text{Thickening(I)} = I \cup H \qquad (8.3)$$

In the thickening operation, the origin of the structuring element has to be either zero or empty. The origin of the structuring element is applied to every pixel in the image (except the edges of the images). The pixel values in the structuring element are compared to the underlying pixels in the sub-image. If all the values in the structuring element match with the pixel values in the sub-image, then the underlying pixel below the origin is set to 1 (foreground). In all other cases, it remains unchanged. In short, the output of the thickening operation consists of the original image and the foreground pixels that have been identified by the hit-or-miss transformation.

Thinning is the opposite of thickening. Thinning is used to remove selected foreground pixels from the image. Thinning is similar to erosion or opening as the thinning operation will result in the shrinking of foreground pixels. Thinning operation can also be expressed in terms of hit-or-miss transformation. The thinning of image I with the structuring element S can be given by Equation 8.4 where H is the hit-or-miss of image I with S,

$$\text{Thinning}(I) = I - H \tag{8.4}$$

In the thinning operation, the origin of the structuring element has to be either 1 or empty. The origin of the structuring element is applied to every pixel in the image (except the edges of the images). The pixel values in the structuring element are compared to the underlying pixels in the image. If all the values in the structuring element match with the pixel values in the image, then the underlying pixel below the origin is set to 0 (background). In all other cases, it remains unchanged.

Both thickening and thinning operations can be applied repeatedly.

8.8.1 Skeletonization

The process of applying the thinning operation multiple times so that only connected pixels are retained is known as skeletonization. This is a form of erosion where most of the foreground pixels are removed and only pixels with connectivity are retained. As the name suggests, this method can be used to define the skeleton of the object in an image.

The following is the Python function for skeletonization:

```
skimage.morphology.skeletonize(image)


Necessary arguments:
image can be ndarray array of either binary or
boolean type. If the image is binary, foreground
pixels are represented by 1 and background pixels
by 0. If the image is boolean, True represents
foreground while false represents background.

Returns: output as an ndarray containing the skeleton
```

The Python code for skeletonization is given below.

```
from scipy.misc import toimage, fromimage
import Image, numpy
from skimage.morphology import skeletonize

# opening the image and converting it to grayscale
a = Image.open('../figures//steps1.png').
    convert('L')
# converting a to an ndarray and normalizing it
a = fromimage(a)/numpy.max(a)
# performing skeletonization
b = skeletonize(a)
# converting b from an ndarray to an image
c = toimage(b)
# saving the image as skeleton_output.png
# in the folder Figures
c.save('../figures//skeleton_output.png')
```

Figure 8.9(a) is the input image for the skeletonization and Figure 8.9(b) is the output image. Notice that the foreground pixels have shrunk and only the pixels that have connectivity survive the skeletonization process. One of the major uses of skeletonization is in measuring the length of objects. Once the foreground pixels have been shrunk to one pixel width, the length of the object is approximately the number of pixels after skeletonization.

8.9 Summary

- The structuring element is important for most of the binary operations.

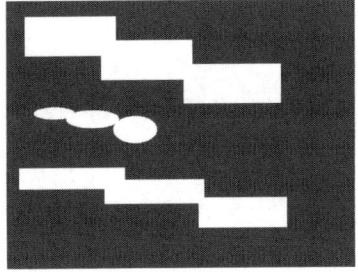 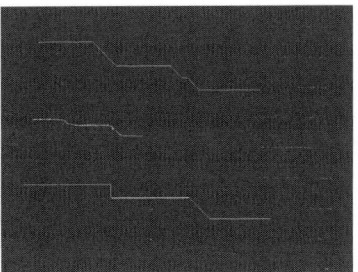

(a) Input image for skeletonization.

(b) Output image after skeletonization.

FIGURE 8.9: An example of skeletonization.

- Binary dilation, closing and thickening operations increase the number of foreground pixels and hence close holes in objects and aggregate nearby objects. The exact effect depends on the structuring element. The closing operation may preserve the size of the object while dilation does not.

- The erosion, opening and thinning operations decrease the number of foreground pixels and hence increase the size of holes in objects and also separate nearby objects. The exact effect depends on the structuring element. The opening operation may preserve the size of the object while erosion does not.

- Hit-or-miss transformation is used to determine specific patterns in an image.

- Skeletonization is a type of thinning operation in which only connected pixels are retained.

8.10 Exercises

1. Perform skeletonization on the image in Figure 8.2(a).

2. Consider an image and prove that erosion followed by dilation is not same as dilation followed by erosion.

3. Imagine an image containing two cells that are next to each other with a few pixels overlapping; what morphological operation would you use to separate them?

4. You are hired as an image processing consultant to design a new checkout machine. You need to determine the length of each vegetable programmatically given an image containing one of the vegetables. Assuming that the vegetables are placed one after the other, what morphological operation will you need?

Chapter 9

Image Measurements

9.1 Introduction

So far we have shown methods to segment an image and obtain various regions that share similar characteristics. An important next step is to understand the shape, size and geometrical characteristics of these regions.

The regions in an image may be circular such as an image of coins or edges in a building. In some cases, the regions may not have simple geometrical shapes like circles, lines etc. Hence radius, slope etc. alone do not suffice to characterize the regions. An array of properties such as area, bounding box, central moments, centroid, eccentricity, euler number etc. are needed to describe shapes of regions.

In this chapter we begin the discussion with a label function that allows numbering each region uniquely, so that the regionprops function can be used to obtain the characteristics. This is followed by Hough transform for characterizing lines and circles. We conclude with a discussion on counting regions or objects using template matching.

9.2 Labeling

Labeling is used to identify different objects in an image. The image has to be segmented before labeling can be performed. In a labeled image, all pixels in a given object have the same value. For example, if an image comprises of four objects, then in the labeled image, all pixels in the first object have a value 1, etc. A labeled image is used as an input image to the regionprops function to determine the properties of the objects.

The Python function for labeling is given below.

```
skimage.morphology.label(image)

Necessary arguments:
    image is the segmented image as an ndarray.

Returns: output labelled image as an ndarray.
```

The Python function for obtaining geometrical characteristics of regions is regionprops. Some of the parameters for regionprops are listed below. The complete list can be found at [87].

```
skimage.measure.regionprops
    (label_image, properties=['Area', 'Centroid'],
                            intensity_image=None)
Necessary arguments:
    label_image is a labelled image as an ndarray.

    output is the Python list of dictionaries.

Optional arguments:
    properties can take the following parameters
(more can be found at the above url):

    Area returns the number of pixels in the object.
It is an integer.
```

BoundingBox returns a tuple consisting of four values:
lower left corner co-ordinates corresponding to the
beginning of the bounding box. Upper right corner
co-ordinates corresponding to the end of the
bounding box.

Centroid returns the co-ordinates of the centroid
of the object.

Image returns a sliced binary region image whose
dimensions are same as the size of the bounding box.

FilledImage returns a binary region image with
filled holes whose dimensions are same as the
size of the bounding box.

Instead of a list of descriptors, the properties can
also be ``all''. In such case, all the properties listed
in the url are returned.

The following is the Python code for obtaining the properties of various regions using regionprops. The input image is read and thresholded using Otsu's method. The various objects are labeled using the label function. At the end of this process, all pixels in a given object have the same pixel value. The labeled image is then given as an input to the regionprops function. The regionprops function calculates the area, centroid and bounding box for each of these regions. Finally, the centroid and bounding box are marked on the image using matplotlib functions.

```
import numpy, math
import scipy.misc
```

```
import matplotlib.pyplot as plt
import matplotlib.patches as mpatches
from skimage.morphology import label
from scipy.misc.pilutil import Image
from skimage.measure import regionprops
from skimage.filter.thresholding
    import threshold_otsu

# opening the image and converting it to grayscale
a = Image.open('../Figures/objects.png').
    convert('L')
# a is converted to an ndarray
a = scipy.misc.fromimage(a)
# threshold value is determined by
# using Otsu's method
thresh = threshold_otsu(a)
# the pixels with intensity greater than
# theshold are kept
b = a > thresh
# labelling is performed on b
c = label(b)
# c is converted from an ndarray to an image
c1 = scipy.misc.toimage(c)
# c1 is saved as label_output.png
c1.save('../Figures/label_output.png')
# on the labelled image c, regionprops is performed
d = regionprops(c, properties=['Area',
    'Centroid','BoundingBox'])

# the following command creates an empty plot of
# dimension 6 inch by 6 inch
fig, ax = plt.subplots(ncols=1,nrows=1,
        figsize=(6, 6))
```

```
# plots the label image on the
# previous plot using colormap
ax.imshow(c, cmap='YlOrRd')

for i in d:
    # printing the x and y values of the
    # centroid where centroid[1] is the x value
    # and centroid[0] is the y value
    print i['Centroid'][1],i['Centroid'][0]
    # plot a red circle at the centroid, ro stands
    # for red
    plt.plot(i['Centroid'][1],i['Centroid'][0],'ro')
    # In the bounding box, (lr,lc) are the
    # co-ordinates of the lower left corner and
    # (ur,uc) are the co-ordinates
    # of the top right corner
    lr, lc, ur, uc = i['BoundingBox']
    # the width and the height of the bounding box
    # is computed
    rec_width = uc - lc
    rec_height = ur - lr

    # Rectangular boxes with
# origin at (lr,lc) are drawn
    rect = mpatches.Rectangle((lc, lr),rec_width,
            rec_height,fill=False,edgecolor='black',
            linewidth=2)
    # this adds the rectangular boxes to the plot
    ax.add_patch(rect)

# displays the plot
plt.show()
```

Figure 9.1(a) is the input image for the regionprops and Figure 9.1(b) is the output image. The output image is labeled with different colors and enclosed in a bounding box obtained using regionprops.

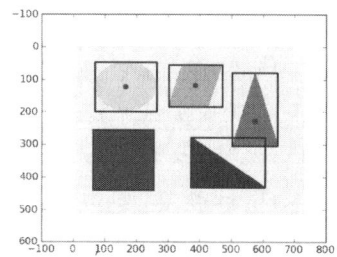

(a) Input image for regionprops.

(b) Labeled output image with bounding boxes and centorids.

FIGURE 9.1: An example of regionprops.

9.3 Hough Transform

The edge detection process discussed in Chapter 4, Spatial Filters, detects edges in an image but does not characterize the slope and intercept of the line or the radius of a circle. These characteristics can be calculated using Hough transform.

9.3.1 Hough Line

The general form of a line is given by $y = mx + b$ where m represents slope of the line and b represents the y-intercept. But in the case of a vertical line m is undefined or infinity and hence the accumulator plane (discussed below) will have infinite length, which cannot be programmed in a computer. Hence, we use polar coordinates which are finite for all slopes and intercepts to characterize a line.

The polar form of a line (also called normal form) is given by the following equation:

$$x\cos(\theta) + y\sin(\theta) = r \tag{9.1}$$

where r is positive and is the perpendicular distance between the origin and the line and θ is the slope of the line and it ranges from $[0, 180]$. Each point in the (x, y) plane also known as the cartesian plane can be transformed into (r, θ) plane also known as the accumulator plane, a 2D matrix.

A segmented image is given as an input for the Hough line transform. To characterize the line, a 2D accumulator plane with r and θ is generated. For a specific (r, θ) and for each x value in the image, the corresponding y value is computed using Equation 9.1. For every y value that is the foreground pixel i.e., the y value lies on the line, a value of 1 is added to the specific (r, θ) in the accumulator plane. This process is repeated for all values of (r, θ). The resultant accumulator plane will have high intensity at the points corresponding to a line. Then the (r, θ) corresponding to the local peak will provide the parameters of the line in the original image.

If the input image is of size N-by-N, the number of values of r is M and number of points in θ is K, the computational time for accumulator array is $O(KMN^2)$. Hence, Hough line transform is a computationally intensive process. If θ ranges from $[0, 180]$ and for a step size of 1, then $K = 180$ along the θ axis. If the range of θ is known a priori and is smaller than $[0, 180]$, K will be smaller and hence the computation can be made faster. Similarly, if other factors such as M or N can be reduced, the computational time can be reduced as well.

The cv2 function for Hough line transform is given below:

```
cv2.HoughLines(image,rho,theta,threshold)
```

Necessary argument:

image should be binary.

rho is the resolution of the distance,
r in pixels.

theta is the resolution of the angle in pixels.

threshold is the minimum value that will be used
to detect a line in the accumulator matrix.

Returns: Outputs is a vector with distance and
    angle of detected lines.

The cv2 code for Hough line transform is given below. The input image (Figure 9.2(a)) is converted to grayscale. The image is then thresholded using Otsu's method (Figure 9.2(b)) to obtain a binary image. On the thresholded image, Hough line transformation is performed. The output of Hough line transform with the detected lines is shown in Figure 9.2(c). The thick lines are lines that are detected by Hough line transform.

```
import numpy as np
import scipy.misc, cv2

# opening the image
im = cv2.imread('../Figures/hlines2.png')
# converting the image to grayscale
a1 = cv2.cvtColor(im,cv2.COLOR_BGR2GRAY)
# thresholding the image to obtain
# only foreground pixels
thresh,b1 = cv2.threshold(a1, 0, 255,
            cv2.THRESH_BINARY_INV+cv2.THRESH_OTSU)
# converting the thresholded ndarray to an image
```

```
b2 = scipy.misc.toimage(b1)
b2.save('../Figures/hlines_thresh.png')
# performing the Hough lines transform
lines = cv2.HoughLines(b1,5,0.1,200)
# printing the lines: distance and angle in radians
print lines
```

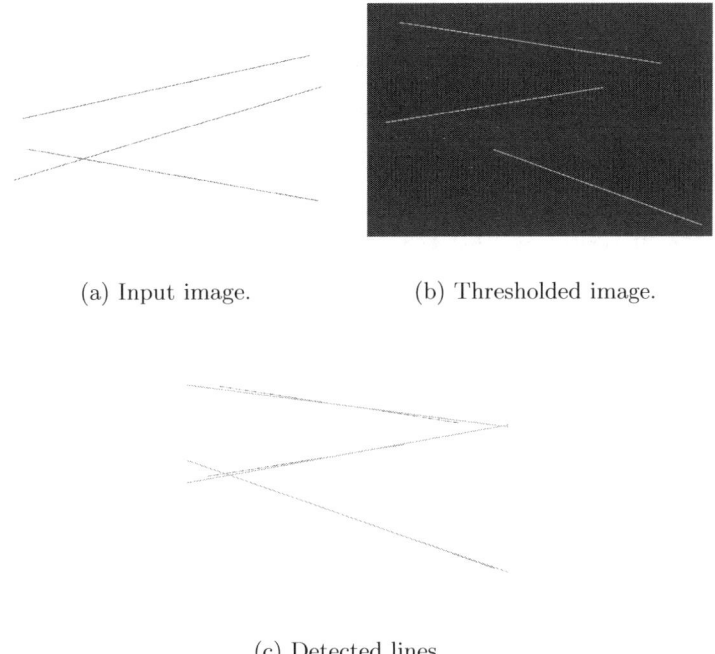

(a) Input image. (b) Thresholded image.

(c) Detected lines.

FIGURE 9.2: An example of Hough line transform.

9.3.2 Hough Circle

The general form of a circle is given by $(x - a)^2 + (y - b)^2 = R^2$ where (a, b) is the center of the circle and R is the radius of the circle. The equation can be rewritten as $y = b \pm \sqrt{R^2 - (x - a)^2}$. Alternately, it can be written in a simpler polar form as

$$\begin{cases} x = a + R\cos(\theta) \\ y = b + R\sin(\theta) \end{cases} \quad (9.2)$$

where θ ranges from $[0, 360]$.

It can be seen from Equation 9.2 that each point in the (x, y) plane can be transformed into (a, b, R) hyper-plane or accumulator plane.

To characterize the circle, a 3D accumulator plane with R, a and b is generated. For a specific (R, a, b) and for each θ value, the corresponding x and y value is computed using Equation 9.2. For every x and y value that is the foreground pixel i.e., the (x, y) value lies on the circle, a value of 1 is added to the specific (R, a, b) in the accumulator plane. This process is repeated for all values of (R, a, b). The resultant accumulator hyper-plane will have high intensity at the points corresponding to a circle. Then the (R, a, b) corresponding to the local peak will provide the parameters of the circle in the original image.

The following is the Python function for Hough circle transform:

```
cv2.HoughCircles(input, CV_HOUGH_GRADIENT, dp,
    min_dist, param1, param2, minRadius, maxRadius);
```

Necessary argument:
input is a grayscale image.

CV_HOUGH_GRADIENT is the method that is used by OpenCV.

dp is the inverse ratio of resolution.

min_dist is the minimum distance that the function will maintain between the detected centers.

param1 is the upper threshold for Canny edge detector

that is used by the Hough function internally.

param2 is the threshold for center detection.

Optional arguments:
  min_radius is the minimum radius of the circle
that needs to be detected while max_radius is
the maximum radius.

Returns: output is a vector that contains information about
the (x,y) values of the center and radius of
each detected circle.

The cv2 code for the Hough circle transform is given below.

```
import numpy as np
import scipy.ndimage
from PIL import Image
import scipy.misc, cv2


# opening the image and converting it to grayscale
a = Image.open('../Figures/withcontrast1.png').
    convert('L')
a = scipy.misc.fromimage(a)
# median filter is performed on the
# image to remove noise
img = scipy.ndimage.filters.median_filter(a,size=5)
# circles are determined using
# Hough circles transform
circles = cv2.HoughCircles(img,
          cv2.cv.CV_HOUGH_GRADIENT,1,10,param1=100,
          param2=30,minRadius=10,maxRadius=30)
# circles is arounded to unsigned integer 16
```

```
circles = np.uint16(np.around(circles))
# For each detected circle
for i in circles[0,:]:
# an outer circle is drawn for visualization
    cv2.circle(img,(i[0],i[1]),i[2],(0,255,0),2)
# its center is marked
    cv2.circle(img,(i[0],i[1]),2,(0,0,255),3)


# converting img from an ndarray to an image
cimg = scipy.misc.toimage(img)
# saving the image as houghcircles_output.png
cimg.save('../Figures/houghcircles_output.png')
```

Figure 9.3(a) is a CT image with two bright white circular regions being contrast-filled blood vessels. The aim of this exercise is to characterize the vessel size using Hough circle transform. The image is median filtered (Figure 9.3(b)) to remove noise. Finally, the output of Hough circle transform in a search space of minimum radius 10 and maximum radius 30 is given in Figure 9.3(c). The two circles that are detected are marked using dark circles.

If the input image is of size N-by-N, the number of values of a and b are M and number of points in R is K, the computational time is $O(KM^2N^2)$. Hence, Hough circle transform is significantly computationally intensive compared to Hough line transform. If the range of radius to be tested is smaller, then K is smaller and hence the computation can be made faster. If the approximate location of the circle is known, then the range of a and b is reduced and consequently decreases M and hence computation can be accomplished faster. Interested readers can refer to [42],[41],[52],[96] and [107] to learn more about Hough transforms.

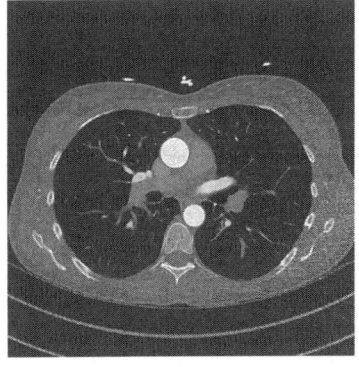 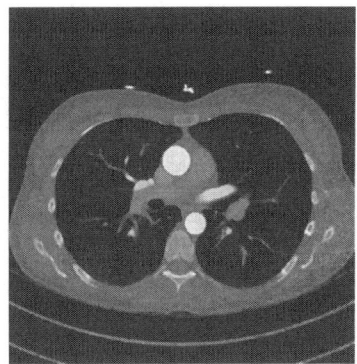

(a) Input Image.   (b) Image after performing median filter.

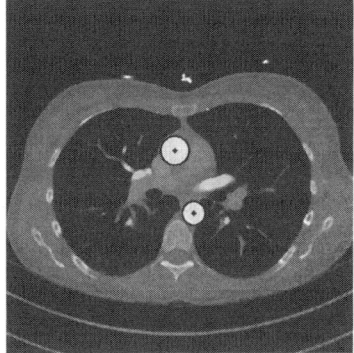

(c) Output with min radius = 10 and max radius 30.

FIGURE 9.3: An example of Hough circle transform.

9.4 Template Matching

Template matching technique is used to find places in an image that match with the given template. Template matching is used to identify a particular person in a crowd or a particular car in traffic etc. It works by comparing a sub-image of the person or object over a much larger image.

Template matching can be either intensity based or feature based. We will demonstrate intensity based template matching. A mathematical coefficient called cross-correlation is used for intensity based template matching. Let $I(x, y)$ be the pixel intensity of image I at (x, y) then the cross-correlation, c between $I(x, y)$ and template $t(u, v)$ is given by

$$c(u, v) = \sum_{x,y} I(x, y) t(x - u, y - v) \qquad (9.3)$$

Cross-correlation is similar to convolution operation. Since $c(u, v)$ is not independent of the changes in image intensities, we use the normalized cross-correlation coefficient proposed by J.P. Lewis [51]. The normalized cross-correlation coefficient is given by the following equation:

$$r(u, v) = \frac{\sum_{x,y}(I(x,y) - \bar{I})(t(x-u, y-v) - \bar{t})}{\sqrt{\sum_{x,y}(I(x,y) - \bar{I})^2 \sum_{x,y}(t(x-u, y-v) - \bar{t})^2}} \qquad (9.4)$$

where $\bar{I}$ is the mean of the sub-image that is considered for template matching and $\bar{t}$ is the average of the template image. In the places where the template matches the image, the normalized cross-correlated coefficient is close to 1.

The following is the Python code for template matching.

```
import scipy.misc
import numpy as np
from skimage import filter
import matplotlib.pyplot as plt
from scipy.misc.pilutil import Image
from skimage.morphology  import label
```

```
from skimage.measure import regionprops
from skimage.feature import match_template

# opening the image and converting it to grayscale
image =Image.open('../Figures/airline_seating.png').
       convert('L')
# converting the input image into an ndarray
image = scipy.misc.fromimage(image)
# reading the template image
temp = Image.open('../Figures/template1.png').
       convert('L')
# converting the template into an ndarray
temp = scipy.misc.fromimage(temp)
# performing template matching
result = match_template(image, temp)
thresh = 0.7
# thresholding the result from template
# matching considering pixel values where the
# normalized cross-correlation is greater than 0.7
res = result > thresh
# labeling the thresholded image
c = label(res, background = 0)
# performing regionprops to count the
# number of labels
reprop = regionprops(c)
print "The number of seats are:", len(reprop)
# converting the ndarray to image
d = scipy.misc.toimage(res)
d.show()
```

The results of template matching are shown in Figure 9.4. Figure 9.4(a) is the input image containing the layout of airline seats and

Figure 9.4(b) is the template image. The normalized cross-correlation coefficient, r, is computed for every pixel in the input image. Then the array comprising of the normalized cross-correlated coefficients is thresholded. The threshold value of 0.7 is chosen. Then the regions in the thresholded array are labeled. Regionprops is performed on the labeled array to obtain the number of regions that match the template and have $r > 0.7$. The output image in Figure 9.4(c) is the thresholded image. The number of seats returned by the program is 263.

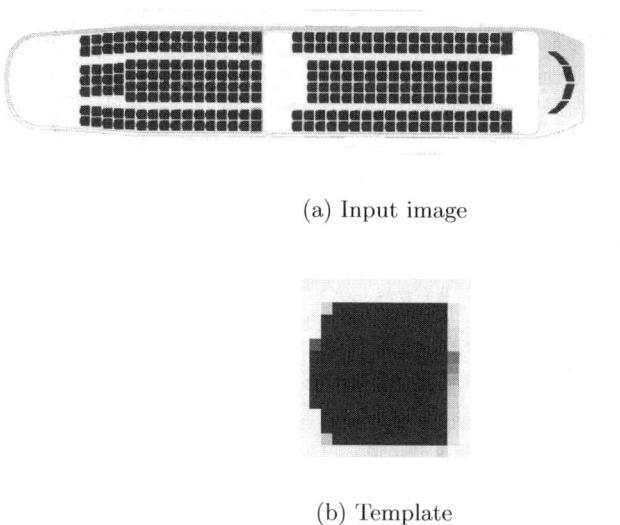

(a) Input image

(b) Template

(c) Cross-correlated image after segmentation

FIGURE 9.4: An example of template matching.

9.5 Summary

- Labeling is used to identify different objects in an image.

- The regionprops function has several attributes and is used to study different properties of objects in a labeled image.

- Hough line transform detects lines while Hough circle transform detects circles. They also determine the corresponding parameters: slope and intercept for lines, and center and diameter for circles.

- Template matching is used to identify or count similar objects in an image.

9.6 Exercises

1. Hough transform is one method for finding the diameter of a circle. The process of finding the diameter is slow. Suggest a method for determining the **approximate diameter** of a circle, given only pixels corresponding to the two blood vessels in Figure 9.3(a).

2. Figure 4.9(a) in Chapter 4 consists of multiple characters. Write a Python program to separate each of these text and store the individual characters as separate images. Hint: Use the regionprops function.

3. Consider an image with 100 coins of various sizes spread on a uniform background. Assume that the coins do not touch each other, write a pseudo code to determine the number of coins for

each size. Brave soul: Write a Python program to accomplish the same. Hint: regionprops will be needed.

4. Consider an image with 100 coins of various sizes spread on a uniform background. Assume that the coins **do touch each other**, and write a pseudo code to plot a histogram of the area of the coin (along the x-axis) vs the number of coins for a given area (along the y-axis). Write a Python program to accomplish the same. If only few coins overlap, determine the approximate number of coins.

Part III

Image Acquisition

Chapter 10

X-Ray and Computed Tomography

10.1 Introduction

So far we have covered the basics of Python and its scientific modules; and image processing techniques. In this chapter, we begin our journey of learning image acquisition. In this chapter, we begin the discussion with x-ray generation and detection. We discuss the various modes in which x-ray interacts with matter. These methods of interaction and detection have resulted in many modes of x-ray imaging such as angiography, fluoroscopy etc. We complete the discussion with the basics of CT, reconstruction and artifact removal.

10.2 History

X-rays were discovered by Wilhelm Conrad Röntgen, a German physicist, during his experiment with cathode ray tubes. He called these mysterious rays "x-rays," the symbol "x" being used in mathematics to denote unknown variables. He found that unlike visible light, these rays passed through most of the materials and left a characteristic shadow on a photographic plate. His work was published as "On A New Kind of Rays" and was subsequently awarded the first Nobel Prize in Physics in 1901. Röntgen's paper can be accessed at the Wiley online library [82].

Subsequent study of x-rays revealed their true physical nature. They are a form of electromagnetic radiation similar to light, radio waves etc. They have a wavelength of 10 to 0.01 nanometers. Although they are well known and studied and no longer mysterious, they continue to be referred to as x-rays. Even though the majority of x-rays are man-made using x-ray tubes, they are also found in nature. The branch of x-ray astronomy studies celestial objects by measuring the x-rays emitted.

Since Röntgen's days, the x-ray has found a very widespread use across various fields including radiology, geology, crystallography, astronomy etc. In the field of radiology, x-rays are used in fluoroscopy, angiography, computed tomography (CT) etc. Today, many non-invasive surgeries are performed under x-ray guidance, providing a new "eye" to the surgeons.

10.3 X-Ray Generation

In principle, an x-ray tube is very simple system. It consists of a generator producing a constant and reliable output of x-rays, an object through which the x-ray traverses and an x-ray detector to measure the intensity of the rays after passing through the object. We begin with a discussion of the x-ray generation process using an x-ray tube.

10.3.1 X-Ray Tube Construction

An x-ray tube consists of four major parts. They are an anode, a cathode, a tungsten target and an evacuated tube to hold the three parts together, as shown in Figure 10.1.

The cathode (negative terminal) produces electrons (negatively charged) that are accelerated towards the anode (positive terminal). The filament is heated by passing current, which generates electrons by a process of thermionic emission. It is defined as emission of electrons

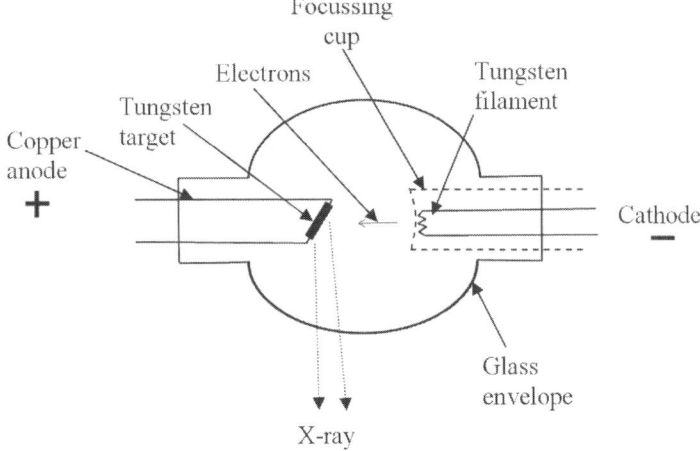

FIGURE 10.1: Components of an x-ray tube.

by absorption of thermal energy. The number of electrons produced is proportional to the current through the filament. This current is generally referred to as "tube current" and is generally measured in "mA" or "milli-amperes".

Since the interior of an x-ray tube can be hot, a metal with a high melting point such as tungsten is chosen for the filament. Tungsten is also a malleable material, ideal for making fine filaments. The electron produced is focused by the focusing cup, which is maintained at the same negative potential as the cathode. The glass enclosure in which the x-ray is generated is evacuated so that the electrons do not interact with other molecules and can also be controlled independently and precisely. The focusing cup is maintained at a very high potential in order to accelerate the electrons produced by the filament.

The anode is bombarded by the fast moving electrons. It is generally made from copper so that the heat produced by the bombardment of the electrons can be properly dissipated. A tungsten target is fixed to the anode. The fast moving electrons either knock out the electrons from the inner shells of the tungsten target or are slowed due to the tungsten nucleus. The former results in the characteristic x-ray spectrum while the latter results in the general spectrum or Bremsstrahlung spectrum.

The two spectrums together determine the energy distribution in an x-ray and will be discussed in detail in the next section.

The cathode is stationary but the anode can be stationary or rotating. The rotating anode allows even distribution of heat and consequently longer life of the x-ray tube.

There are three parameters that control the quality and quantity of an x-ray. These parameters together are sometimes referred to as x-ray technique.

They are:

1. Tube voltage measured in kVp

2. Tube current measured in mA

3. X-ray exposure time in ms

In addition, a filter (such as a sheet of aluminum) is placed in the path of the beam, so that lower energy x-rays are absorbed. This will be discussed in the next section.

The tube voltage is the electric potential between the cathode and the anode. Higher voltage results in increased velocity of the electrons between the cathode and the anode. This increased velocity will produce high energy x-rays will be discussed in subsequent sections. Lower voltage results in lower energy x-rays and consequently noisier image. The tube current determines the number of electrons being emitted. This in turn determines the quantity of x-rays. The exposure time determines the time for which the object or patient is exposed to x-rays. This is generally the time the x-ray tube is operating.

10.3.2 X-Ray Generation Process

The x-ray generated by the tube does not contain photons of single energy. It instead consists of a large range of energy. The relative number of photons at each energy level is measured to generate a histogram. This histogram is called the spectral distribution or spectrum for short. There are two types of x-ray spectrums [16]. They are the general radiation or Bremsstrahlung "Braking" spectrum which is a continuous radiation, and the characteristic spectrum, a discrete entity as shown in the Figure 10.2.

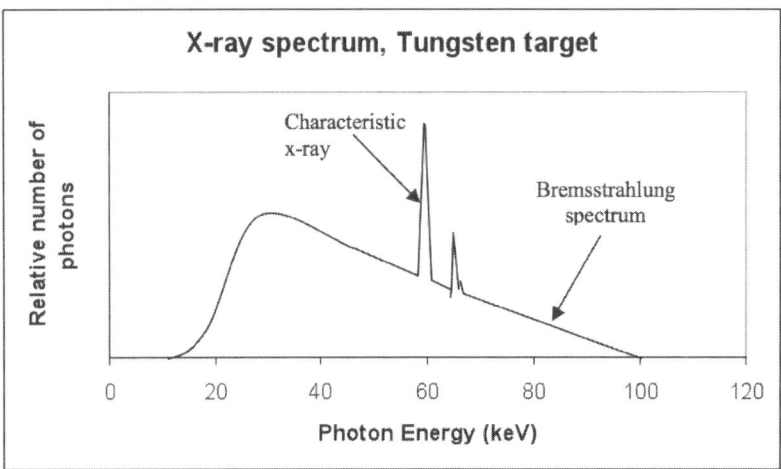

FIGURE 10.2: X-ray spectrum illustrating characteristic and Bremsstrahlung spectrum.

When the fast moving electrons produced by the cathode move very close to the nucleus of the tungsten atom (Figure 10.3), the electrons decelerate and the loss of energy is emitted as radiation. Most of the radiation is at a higher wavelength (or lower energy) and hence is dissipated as heat. The electrons are not decelerated completely by one tungsten nucleus and hence at every stage of deceleration, radiation of lower wavelength or higher energy is emitted. Since the electrons are decelerated or "braked" in the process, this spectrum is referred to as Bremsstrahlung or braking spectrum. This spectrum gives the x-ray spectrum its wide range of photon energy levels.

From the energy equation, we know that

$$E = \frac{hc}{\lambda} \qquad (10.1)$$

where $h = 4.135 * 10^{-18} eVs$ is the Planck's constant, $c = 3 * 10^8 m/s$

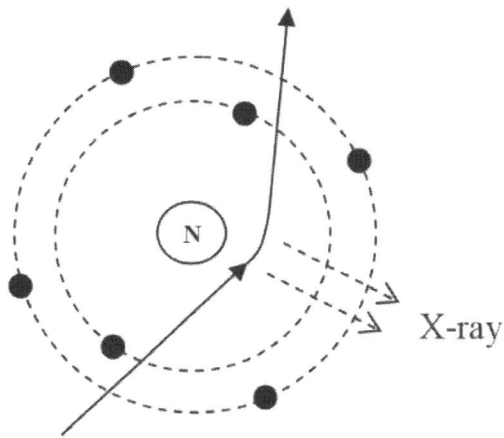

FIGURE 10.3: Production of Bremsstrahlung or braking spectrum.

is the speed of light and λ is the wavelength of the x-ray measured in Armstrong ($A^0 = 10^{-10}m$). The product of h and c is $12.4*10^{-10} keV m$. When E is measured in keV, the equation simplifies to

$$E = \frac{12.4}{\lambda} \qquad (10.2)$$

The inverse relationship between E and λ implies that a shorter wavelength produces a higher energy x-ray and vice-versa. For a x-ray tube powered at 112 kVp, the maximum energy that can be produced is 112 keV and hence the corresponding wavelength is 0.11 A^0. This is the shortest wavelength and also the highest energy that can be achieved during the production of Bremsstrahlung spectrum. This is the right most point in the graph in figure 10.2. However, most of the x-ray will be produced at much higher wavelength and consequently lower energy.

The second type of radiation spectrum (Figure 10.4) results from a tungsten electron in its orbit interacting with the emitted electron. This is referred to as characteristic radiation, as the histogram of spectrum is a characteristic of the target material.

The fast moving electrons eject the electron from the k-shell (inner shell) of the tungsten atom. Since this shell is unstable due to the ejection of the electron, the vacancy is filled by electron from the outer shell. This is accompanied by release of x-ray energy. The energy and wavelength of the electron are dependent on the binding energy of the electron whose position is filled. Depending on the shell, these characteristic radiations are referred as K, L, M and N characteristic radiation and are shown in Figure 10.2.

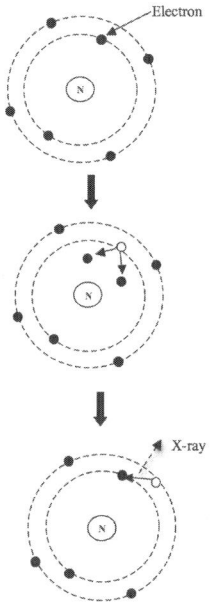

FIGURE 10.4: Production of characteristic radiation.

X-rays do not just interact with the tungsten atom, they can interact with any atom in their path. Thus, a molecule of oxygen in the path will be ionized by an x-ray knocking out its electron. This could change the x-ray spectrum and hence the x-ray generator tube is maintained at vacuum.

10.4 Material Properties

10.4.1 Attenuation

Once the x-ray is generated, it is allowed to pass through a patient or an object. The material in the object reduces the intensity of the x-ray. This process is referred to as attenuation. It can be defined as the reduction in the intensity of the x-ray beam as it traverses matter by either the absorption or deflection of photons in the beam. If there are multiple materials, each of the materials can absorb or deflect the x-ray and consequently reduce its intensity.

The attenuation is quantified by using linear attenuation coefficient (μ), defined as the attenuation per centimeter of the object. The attenuation is directly proportional to the distance traveled and the incident intensity. The intensity of the x-ray beam after attenuation is given by the Lambert Beer law (Figure 10.5) expressed as

$$I = I_0 e^{-\mu \delta x} \qquad (10.3)$$

where I_0 is the initial x-ray intensity, I is the exiting x-ray intensity, μ is the linear attenuation coefficient of the material, and δx is the thickness of the material. The law also assumes that the input x-ray intensity is mono-energetic or monochromatic.

Monochromatic radiation is characterized by photons of single intensity, but in reality all radiations are polychromatic and have photons of varying intensity with a spectra similar to Figure 10.2. Polychromatic radiation is characterized by photons of varying energy (quality and quantity), with the peak energy being determined by the peak kilovoltage (kVp).

When polychromatic radiation passes through matter, the longer wavelengths and lower energy are preferentially absorbed. This increases the mean energy of the beam. This process of increased mean energy of the beam is referred to as "beam hardening".

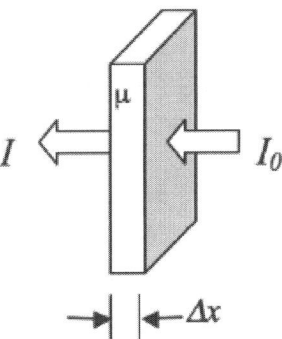

FIGURE 10.5: Lambert Beer law for monochromatic radiation and for a single material.

In addition to the attenuation coefficient, the characteristics of a material under x-ray can also be defined using the half-value layer. This is defined as the thickness of material needed to reduce the intensity of the x-ray beam by half. So from Equation 10.3 for a thickness $\delta x = HVL$ (half value layer),

$$I = \frac{I_0}{2} \qquad (10.4)$$

Hence,

$$I_0 e^{-\mu HVL} = \frac{I_0}{2} \qquad (10.5)$$

$$\mu HVL = 0.693 \qquad (10.6)$$

$$HVL = \frac{0.693}{\mu} \qquad (10.7)$$

For a material with linear attenuation coefficient of 0.1/cm, the HVL is 6.93 cm. This implies that when a monochromatic beam of x-ray passes through the material, its intensity drops by half after passing through 6.93 cm of that material.

kVp	HVL(mm of Al)
50	1.9
75	2.8
100	3.7
125	4.6
150	5.4

TABLE 10.1: Relationship between kVp and HVL.

The HVL depends not only on the material being studied but also on the tube voltage. High tube voltage produces smaller number of low energy photons i.e., the spectrum in Figure 10.2 will be shifted to the right. The mean energy will be higher and the beam will be harder. This hardened beam can penetrate material without a significant loss of energy. Thus, HVL will be high for high x-ray tube voltage. This trend can be seen in the HVL of aluminum at different tube voltages given in Table 10.1.

10.4.2 Lambert Beer Law for Multiple Materials

For an object with n elements (Figure 10.6), the Lambert-Beer law is applied in cascade,

$$I = I_0 e^{-\mu_1 \Delta x} e^{-\mu_2 \delta x} ... e^{-\mu_n \delta x} = I_0 e^{-\sum_{i=1}^{n} \mu_i \Delta x} \tag{10.8}$$

When the logarithm of the intensities is taken, for a continuous domain we obtain

$$p = -\ln\left(\frac{I}{I_0}\right) = \sum_{i=1}^{n} \mu_i \Delta x = \int \mu(x) dx \tag{10.9}$$

Using this equation, we see that the value p, the projection image expressed in energy intensity, corresponding to the digital value at a specific location in that image, is simply the sum of the product of attenuation coefficients and thicknesses of the individual components.

This is the basis of image formation in x-ray and CT that will be discussed shortly.

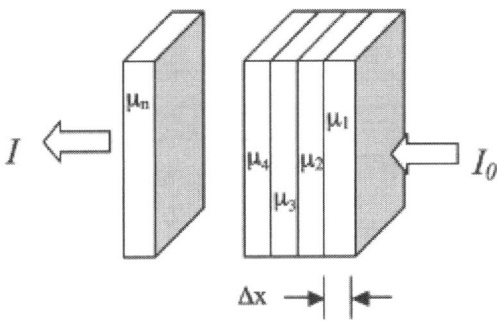

FIGURE 10.6: Lambert Beer law for multiple materials.

10.5 X-Ray Detection

So far, we have discussed the x-ray generation using an x-ray tube, shape of x-ray spectrum and also studied the change in x-ray intensity as it traverses a material due to attenuation. These attenuated x-rays have to be converted to a human viewable form. This conversion process can be achieved either by exposing them on a photographic plate to obtain an x-ray image or viewing them using a TV screen or converting to a digital image, all using the process of x-ray detection. There are three different types of x-ray radiation detectors in practice, namely ionization, fluorescence and absorption.

1. Ionization detection

    In the ionization detector, the x-rays ionize the gas molecules in the detector and by measuring the ionization, the intensity of the x-ray is measured. Example of such a detector is the Geiger Muller counter [55] shown in Figure 10.7. These detectors are used to measure the intensity of radiation and are not used for creating x-ray images.

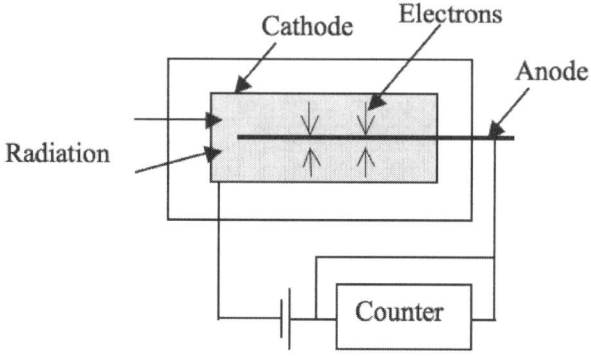

FIGURE 10.7: Ionization detector.

2. Scintillation detection

   There are different types of scintillation detectors. The most popular are Image Intensifier (II) and Flat Panel Detector (FPD). In II, [55], [16], [24], the x-rays are converted to electrons that are accelerated to increase their energy. The electrons are then converted back to light and are viewed on a TV or a computer screen. In the case of FPD, the x-rays is converted to visible light and then to electrons using photo diode. The electrons are recorded using a camera. In both II and FPD, the process of converting x-ray is used for improving the image gain. Modern technology has allowed the creation of large FPD with very high quality and hence FPD is rapidly replacing II. Also, FPD occupies significantly less space than II. We will discuss each of these in detail.

10.5.1 Image Intensifier

The II (Figure 10.8) consists of an input phosphor and photocathode, an electrostatic focusing lens, an accelerating anode and an output fluorescent screen. The x-ray beam passes through the patient and enters the II through the input phosphor. The phosphor generates light photons after absorbing the x-ray photons. The light photons are absorbed by the photocathode and electrons are emitted. The electrons are then accelerated by a potential difference towards the anode. The anode focuses the electron onto an output fluorescence screen that emits the light that will be displayed using a TV screen, recorded on an x-ray film, or recorded by a camera onto a computer.

The input phosphor is made of cesium iodide (CsI) and is vapor deposited to form a needlelike structure that prevents diffusion of light and hence improves resolution. It also has greater packing density and hence higher conversion efficiency even with smaller thickness (needed for good spatial resolution). A photocathode emits electrons when light photons are incident on it. The anode accelerates the electrons. The higher the acceleration the better is the conversion of electrons to light photons at the output phosphor. The input phosphor is curved, so that electrons travel the same length towards the output phosphor. The output fluorescent screen is silver-activated zinc-cadmium sulfide. The output can be viewed using a series of lenses on a TV or it can be recorded on a film.

10.5.2 Multiple-Field II

The field size is changed by changing the position of the focal point, the point of intersection for the left and right electron beams. This is achieved by increasing the potential in the electrostatic lens. Lower potential results in the focus being close to the anode and hence the full view of the anatomy is exposed to the output phosphor. At higher potential, the focus moves away from the anode and hence only a portion of the input phosphor is exposed to the output phosphor. In both cases, the size of the input and output phosphor remains the same but in the smaller mode, a portion of the image from the input phosphor is removed from the view due to a farther focal point.

In a commercial x-ray unit, these sizes are specified in inches. A

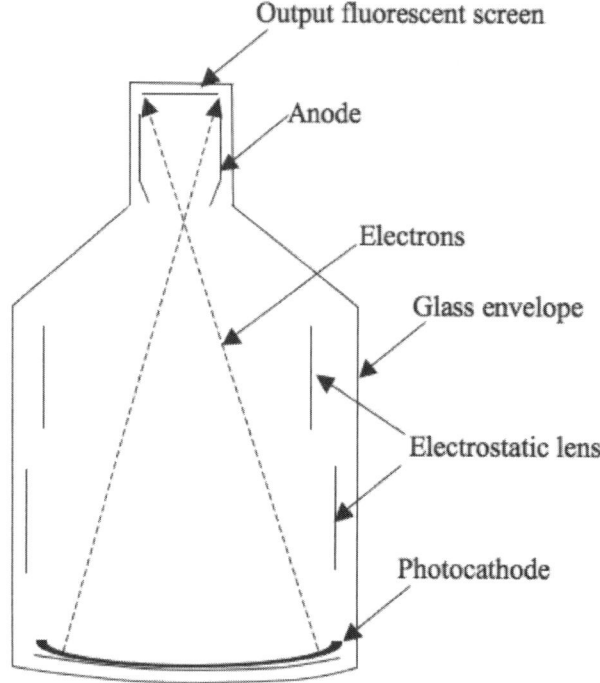

FIGURE 10.8: Components of an image intensifier.

12-inch mode will cover a larger anatomy while a 6-inch mode will cover a smaller anatomy. Exposure factors are automatically increased for smaller II modes to compensate for the decreased brightness from minification.

Since the electrons travel large distances during their journey from photocathode to anode, they are affected by the earth's magnetic field. The earth's magnetic field changes even for small motions of the II and hence the electron path gets distorted. The distorted electron path produces a distorted image on the output fluorescent screen. The distortion is not uniform but increases near the edge of the II. Hence the distortion is more significant for a large II mode than for a smaller II mode. The distortions can be removed by careful design and material selection or more preferably using image processing algorithms.

10.5.3 Flat Panel Detector (FPD)

The FPD (Figure 10.9) consists of a scintillation detector, a photo diode, an amorphous silicon and a camera. The x-ray beam passes through the patient and enters the FPD through the scintillation detector. The detector generates light photons after absorbing the x-ray photons. The light photons are absorbed by the photo diode and electrons are emitted. The electrons are then absorbed by the amorphous silicon layer that produces an image that will be recorded using a charge couple device (CCD) camera.

Similar to II, the scintillation detector is made of cesium iodide (CsI) or gadolinium oxysulfide and is vapor deposited to form needle like structure, which acts like fiber optic cable and prevents diffusion of light and improves resolution. The CsI is generally coupled with amorphous silicon, as CsI is an excellent absorber of x-ray and emits light photons at a wavelength best suited for amorphous silicon to convert to electron.

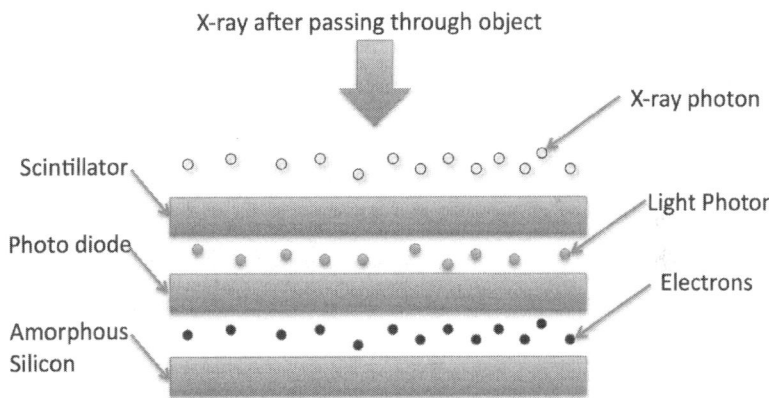

FIGURE 10.9: Flat panel detector schematic.

The II needs extra length to allow accelerating of the electron, while the FPD does not. Hence the FPD occupies significantly less space compared to II. The difference becomes significant as the size of the detector increases. II's are affected by the earth's magnetic field while such problems do not exist for FPD. Hence FPD can be mounted on a x-ray machine and be allowed to rotate around the patient without distorting the image. Although II suffers from some disadvantages, it is simpler in its construction and electronics.

The II or FPD can be bundled with an x-ray tube, a patient table and a structure to hold all these parts together, to create an imaging system. Such a system could also be designed to revolve around the patient table axis and provide images in multiple directions to aid diagnosis or medical intervention. Examples of such systems, fluoroscopy and angiography, are discussed below.

10.6 X-Ray Imaging Modes

10.6.1 Fluoroscopy

The first generation fluoroscope [55],[16] consisted of a fluoroscopic screen made of copper-activated cadmium sulfide that emitted light in the yellow-green spectrum of visible light. The examination was so faint that it was carried out in a dark room, with the doctors adapting their eyes to the dark prior to examination. Since the intensity of fluorescence was less, rod vision in the eye was used and hence the ability to differentiate shades of grey was also poor. These problems were alleviated with the invention of II discussed earlier. The II allowed intensification of the light emitted by the input phosphor so that it could safely and effectively be used to produce a system (Figure 10.10) that could generate and detect x-rays and also produce images that can be studied using TVs and computers.

10.6.2 Angiography

A digital angiographic system [55], [16] consists of an x-ray tube, a detector such as II or FPD and a computer to control the system and record or process the images. The system is similar to fluoroscopy

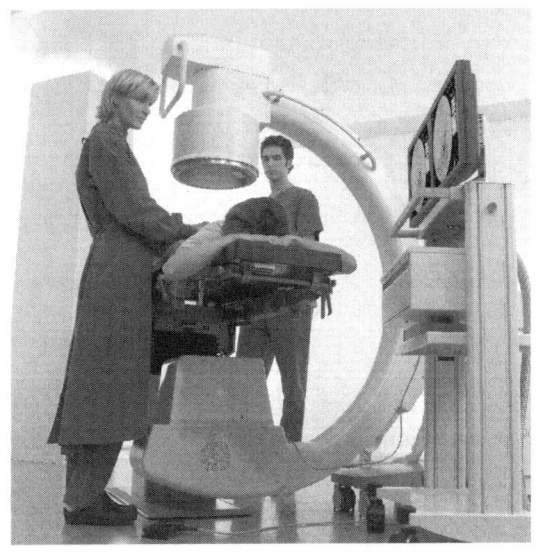

(a) Fluoroscopy machine.

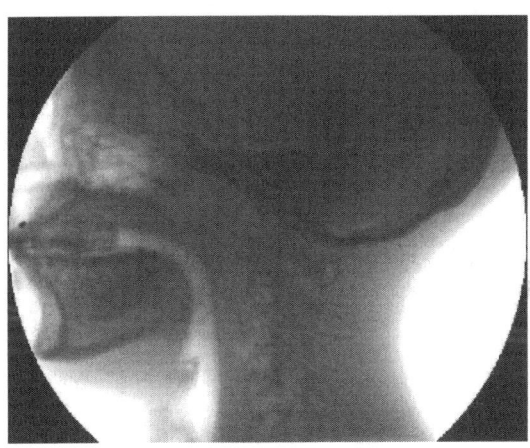

(b) Image of a head phantom acquired using a II system.

FIGURE 10.10: Fluoroscopy machine. Original image reprinted with permission from Siemens AG.

except that it is primarily used to visualize blood vessels opacified using a contrast. The x-ray tube must have a larger focal spot and also provide a constant output over time. The detector must also provide a constant acceleration voltage to prevent variation in gain during acquisition. A computer controls the whole imaging chain and also performs digital subtraction in the case of digital subtraction angiography (DSA) [16] on the obtained images.

In the DSA process, the computer controls the x-ray technique so that uniform exposure is obtained across all images. The computer obtains the first set of images without the injection of contrast and stores them as mask image. Subsequent images obtained under the injection of contrast are stored and subtracted from the mask image to obtain the image with the blood vessel alone.

10.7 Computed Tomography (CT)

The fluoroscopy and angiography discussed so far produce a projection image, which is a shadow of part of the body under x-ray. These systems provide a planar view from one direction and may also contain other organs or structures that impede the ability to make a clear diagnosis. CT on the other hand, provides a slice through the patient and hence offers an unimpeded view of the organ of interest. In CT a series of x-ray images are acquired all around the object or patient. A computer then processes these images to produce a map of the original object using a process called reconstruction. Sir Godfrey N. Hounsfield and Dr. Allan McCormack developed CT independently and later shared the Nobel Prize for Physiology in 1979. The utility of this technique became so apparent that an industry quickly developed around it, and it continues to be an important diagnostic tool for physicians and surgeons. For more details refer to [10],[33],[46].

10.7.1 Reconstruction

The basic principle of reconstruction is that the internal structure of an object can be computed from multiple projections of that object. In the case of CT reconstruction, the internal structure being reconstructed is the spatial distribution of the linear attenuation coefficients (μ) of the imaged object. Mathematically, Equation 10.9 can be inverted by the reconstruction process to obtain the distribution of the attenuation coefficients.

In clinical CT, the raw projection data is often a series of 1D vectors obtained at various angles for which the 2D reconstruction yields a 2D attenuation coefficient matrix. In the case of 3D CT, a series of 2D images obtained at various angles are used to obtain a 3D distribution of the attenuation coefficient. For the sake of simplicity, the reconstructions discussed in this chapter will focus on 2D reconstructions and hence the projection images are 1D vector unless otherwise specified.

10.7.2 Parallel Beam CT

The original method used for acquiring CT data used parallel-beam geometry such as is shown in Figure 10.11. As shown in the figure, the paths of the individual rays of x-ray from the source to the detector are parallel to each other. An x-ray source is collimated to yield a single x-ray beam, and the source and detector is translated along the axis perpendicular to the beam to obtain the projection data (a single 1D vector for a 2D CT slice). After the acquisition of one projection, the source-detector assembly is rotated and subsequent projections are obtained. This process is repeated until a 180 degree projection is obtained. The reconstruction is obtained using the central slice theorem or the Fourier slice theorem [45]. This method forms the basis for many CT reconstruction techniques.

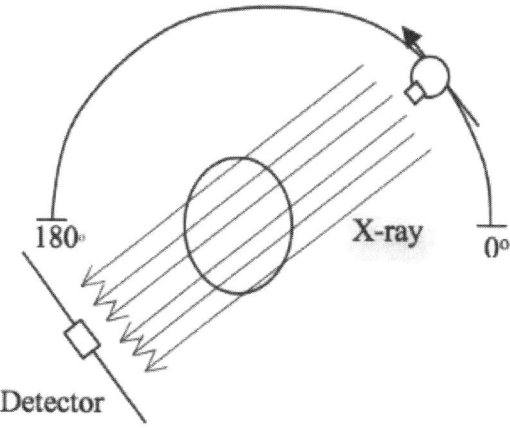

FIGURE 10.11: Parallel beam geometry.

10.7.3 Central Slice Theorem

Consider the distribution or object shown in Figure 10.12 to be reconstructed. The original coordinate system is x-y and when the detector and x-ray source are rotated by an angle θ, then their coordinate system is defined by $x' - y'$. In this figure, R is the distance between the iso-center (i.e., center of rotation) and any ray passing through the object. After logarithmic conversion, the x-ray projection at an angle (θ) is given by

$$g_\theta(R) = \int \int f(x,y)\delta(x\cos\theta + y\sin\theta - R)dx\,dy \quad (10.10)$$

where δ is the Dirac-Delta function [8].

The Fourier transform of the distribution is given by

$$F(u,v) = \int \int f(x,y)e^{-i2\pi(ux+vy)}dx\,dy \quad (10.11)$$

where u and v are frequency components in perpendicular directions. Expressing u and v in polar coordinate, we obtain $u = \nu\cos\theta$ and

$v = \nu \sin\theta$, where ν is the radius and θ is the angular position in the Fourier space.

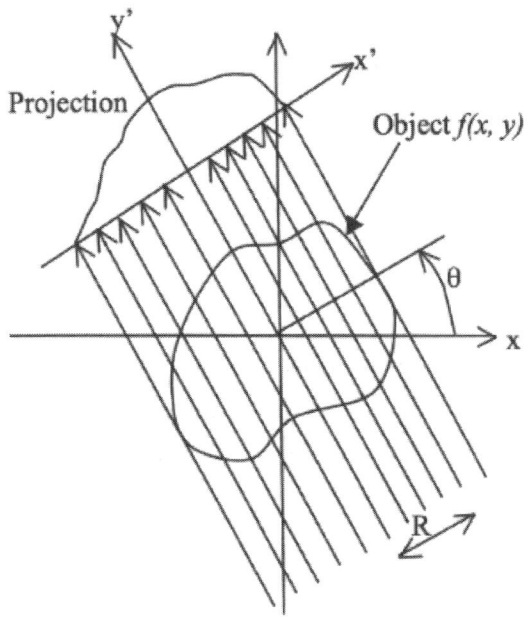

FIGURE 10.12: Central slice theorem.

Substituting for u and v and simplifying yields,

$$F(\nu, \theta) = \int\int f(x,y) e^{-i2v\pi(x\cos\theta + y\sin\theta)} dx\, dy \qquad (10.12)$$

The equation can be rewritten as

$$F(\nu, \theta) = \int\int\int f(x,y) e^{-i2\pi vR} \delta(x\cos\theta + y\sin\theta - R) dR\, dx\, dy \qquad (10.13)$$

Rearranging the integrals yields,

$$F(\nu, \theta) = \int \left(\int\int f(x,y)\delta(x\cos\theta + y\sin\theta - R) \right) e^{-i2\pi vR} dR \qquad (10.14)$$

From Equation 10.10, we can simplify the above equation as

$$F(\nu, \theta) = \int g_e(R) e^{i2\pi \nu R} dR = FT(g_e(R)) \qquad (10.15)$$

where FT( ) refers to the Fourier transform of the enclosed function. Equation 10.15 shows that the radial slice along an angle θ in the 2D Fourier transform of the object is the 1D Fourier transform of the projection data acquired at that angle θ. Thus, by acquiring projections at various angles, the data along the radial lines in the 2D Fourier transform can be obtained. Note that the data in the Fourier space is obtained using polar sampling. Thus, either a polar inverse Fourier transform must be performed or the obtained data must be interpolated onto a rectilinear Cartesian grid so that Fast Fourier Transform (FFT) techniques can be used.

However, another approach can be also taken. Again, $f(x, y)$ is related to the inverse Fourier transform, i.e.,

$$f(x, y) = \int \int F(\nu, \theta) e^{i2\pi(ux+vy)} du\, dv \qquad (10.16)$$

By using a polar coordinate transformation, u, v can be written as $u = \cos\theta$ and $v = \sin\theta$. To effect a coordinate transformation, the Jacobian is used and is given by

$$J = \begin{vmatrix} \frac{\partial u}{\partial \nu} & \frac{\partial u}{\partial \theta} \\ \frac{\partial v}{\partial \nu} & \frac{\partial v}{\partial \theta} \end{vmatrix} = \begin{vmatrix} \cos\theta & -\nu\sin\theta \\ \sin\theta & \nu\cos\theta \end{vmatrix} = \nu \qquad (10.17)$$

Hence,

$$du\, dv = |\nu| d\nu\, d\theta \qquad (10.18)$$

Thus,

$$f(x, y) = \int \int F(\nu, \theta) e^{i2\pi(x\cos\theta + y\sin\theta)} |\nu| d\nu\, d\theta \qquad (10.19)$$

Using Equation 10.15, we can obtain

$$f(x,y) = \int\int FT(g_\theta(R))e^{i2\pi(x\cos\theta+y\sin\theta)}|\nu|d\nu\ d\theta \qquad (10.20)$$

$$f(x,y) = \int\int FT(g_\theta(R))e^{i2\pi\nu R}\delta(x\cos\theta+y\sin\theta-R)|\nu|d\nu\ d\theta\ dR \qquad (10.21)$$

$$f(x,y) = \int\int \left(FT(g_\theta(R))|\nu|e^{i2\pi\nu R}d\nu\right)\delta(x\cos\theta+y\sin\theta-R)d\theta\ dR \qquad (10.22)$$

The term in the braces is the filtered projection, which can be obtained by multiplying the Fourier transform of the projection data by $|\nu|$ in the Fourier space or equivalently by performing a convolution of the real space projections and the inverse Fourier transform of the function $|\nu|$. Because the function looks like a ramp, the filter generated is commonly called the "ramp filter". Thus,

$$f(x,y) = \int\int FT(R,\theta) \bullet \delta(x\cos\theta+y\sin\theta-R)d\theta\ dR \qquad (10.23)$$

where $FT(R,\theta)$ is the filtered projection data at location R acquired at angle θ is given by

$$f(x,y) = \int\int FT(g_\theta(R))|\nu|e^{i2\pi\nu R}dR \qquad (10.24)$$

Once the convolution or filtering is performed, the resulting data is reconstructed using Equation 10.24. This process is referred to as the Filtered back projection (FBP) technique and is the most commonly used technique in practice.

10.7.4 Fan Beam CT

The fan-beam CT scanners (Figure 10.13) have a bank of detectors, with all detectors being illuminated by x-rays simultaneously from every projection angle. Since the detector acquires images in one x-ray exposure, it eliminates the translation at each angle. Since translation is eliminated, the system is mechanically stable and faster. However, x-rays scattered by the object reduce the contrast in the reconstructed images compared to parallel beam reconstruction. But these machines are still popular due to faster acquisition time which allows reconstruction of a moving object, like slices of the heart in one breath-hold. The images acquired using fan beam scanners can be reconstructed using a rebinning method that converts fan beam data into parallel beam data and then uses central slice theorem for reconstruction. Currently, this approach is not used and is replaced by a direct fan beam reconstruction method based on filtered back-projection.

A fan beam detector with one row of detecting elements produces one CT slice. The current generations of fan beam CT machines have multiple detector rows and can acquire 8, 16, 32 slices etc. in one rotation of the object and are referred to as multi-slice CT machines. The benefit is faster acquisition time compared to single slice and also covering a larger area in one exposure. With the advent of multi-slice CT machines, a whole body scan of the patient can also be obtained.

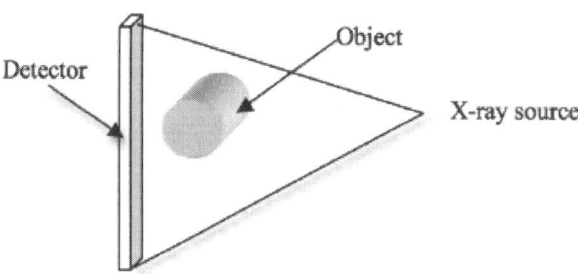

FIGURE 10.13: Fan beam geometry.

Figure 10.14 is the axial slice of the region around the human kidney.

It is one of the many slices of the whole body scan shown in the montage in Figure 10.15. These slices were converted into 3D object (Figure 10.16) using Mimics™ [59].

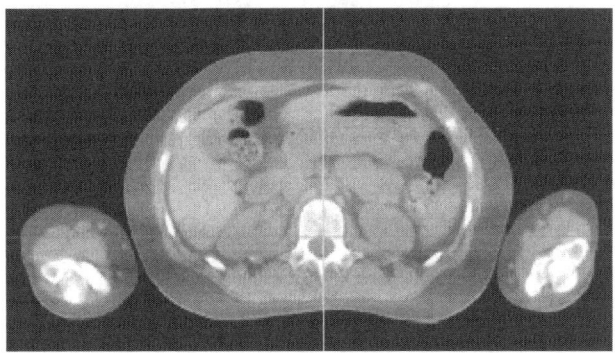

FIGURE 10.14: Axial CT slice.

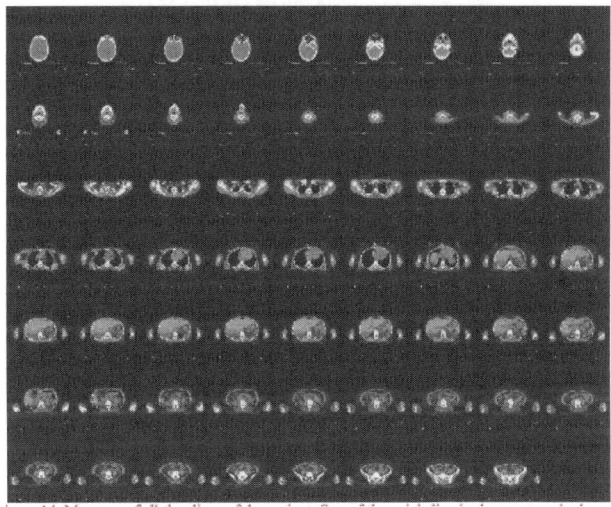

FIGURE 10.15: Montage of all the CT slices of the human kidney region.

10.7.5 Cone Beam CT

Cone beam acquisition or CBCT (Figure 10.17) consists of 2D detectors instead of 1D detectors used in the parallel and fan-beam ac-

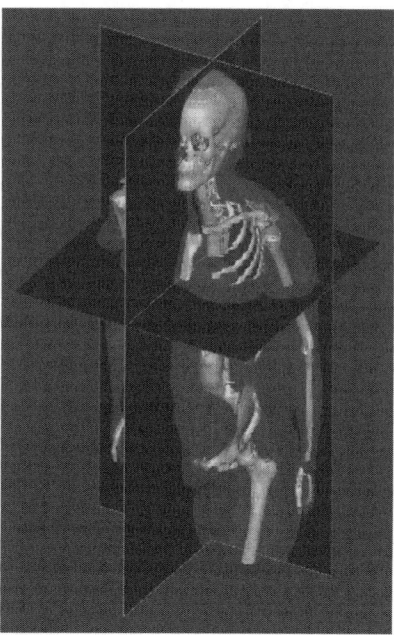

FIGURE 10.16: 3D object created using the axial slices shown in the montage. The 3D object in green is superimposed on the slice information for clarity.

quisitions. As with fan-beam, the source and detector rotate relative to the object, and the projection images are acquired. The 2D projection images are then reconstructed to obtain 3D volume. Since a 2D region is imaged, cone-beam-based volume acquisition makes use of x-rays that otherwise would have been blocked. The advantages are potentially faster acquisition time, better pixel resolution and isotropic (same voxel size in x, y and z directions) voxel resolution. The most commonly used algorithm for cone-beam reconstruction is the Feldkamp algorithm [25], which assumes a circular trajectory for the source and flat detector and is based on filtered backprojection.

10.7.6 Micro-CT

Micro-tomography (commonly known as industrial CT scanning), like tomography, uses x-rays to create cross-sections of a 3D object that

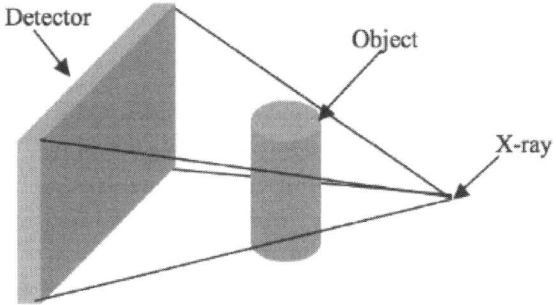

FIGURE 10.17: Cone beam geometry.

later can be used to recreate a virtual model without destroying the original model. The term micro is used to indicate that the pixel sizes of the cross-sections are in the micrometer range. These pixel sizes have also resulted in the terminology micro-computed tomography, micro-CT, micro-computer tomography, high-resolution x-ray tomography, and similar terminologies. All of these names generally represent the same class of instruments.

This also means that the machine is much smaller in design compared to the human version and is used to image smaller objects. In general, there are two types of scanner setups. In the first setup, the x-ray source and detector are typically stationary during the scan while the animal or specimen rotates. In the second setup, much more like a clinical CT scanner, the animal or specimen is stationary while the x-ray tube and detector rotates.

The first x-ray micro-CT system was conceived and built by Jim Elliott in the early 1980s [22]. The first published x-ray micro-CT images were reconstructed slices of a small tropical snail, with pixel size about 50 micrometers appeared in the same paper.

Micro-CT is generally used for studying small objects such as polymers, plastics, micro devices, electronics, paper, fossils. It is also used in the imaging of small animals such as mice, or insects etc.

10.8 Hounsfield Unit (HU)

Hounsfield unit (HU) is a linear transformation that is used to calibrate images acquired using different CT machines. The HU transformation is given by

$$HU = \left(\frac{\mu - \mu_w}{\mu_w}\right) * 1000 \qquad (10.25)$$

where μ is the linear attenuation coefficient of the object and μ_w is the linear attenuation coefficient of water. Thus, water has an HU of 0 and air has an HU of -1000.

The following are the steps to obtain the HU equivalent of a reconstructed image:

- A water phantom consisting of a cylinder filled with water is reconstructed using the same x-ray technique as the reconstructed patient slices.

- The attenuation coefficient of water and air (present outside the cylinder) is measured from the reconstructed slice.

- A linear fit is established with the HU of water (0) and air (-1000) being the ordinate and the corresponding linear attenuation coefficients measured from the reconstructed image being the abscissa.

- Any patient data reconstructed is then mapped to HU using the determined linear fit.

Since the CT data is calibrated to HU, the data in the images acquires meaning not only qualitatively but also quantitatively. Thus, an HU number of 1000 for a given pixel or voxel represents quantitatively a bone in an object.

Unlike MRI, microscopy, ultrasound etc., due to use of HU for calibration, CT measurement is a map of physical property of the material.

This is handy while performing image segmentation, as the same threshold or segmentation technique can be used for measurements from various patients at various intervals and conditions. It is also useful in performing quantitative CT, a process of measuring the property of the object using CT.

10.9 Artifacts

In all the prior discussions, it was assumed that the x-ray beam is mono-energetic. It was also assumed that the geometry of the imaging system is well characterized, i.e. there is no change in the orbit that the imaging system follows with reference to the object. However, in current clinical CT technology, the x-ray beam is not mono-energetic and the geometry is not well characterized. This results in errors in the reconstructed image that are commonly referred as artifacts. An artifact is defined as any discrepancy between the reconstructed value in the image and the true attenuation coefficients of the object [39]. Since the definition is broad and can encompass many things, discussions of artifacts are generally limited to clinically significant errors. CT is more prone to artifacts than conventional radiography, as multiple projection images are used. Hence errors in different projection images cumulate to produce artifacts in the reconstructed image. These artifacts could annoy radiologists or in some severe cases hide important details that could lead to misdiagnosis.

Artifacts can be eliminated to some extent during acquisition. They can also be removed by pre-processing projection images or post-processing the reconstructed images. There are no generalized techniques for removing artifacts and hence new techniques are devised depending on the application, anatomy etc. Artifacts cannot be completely eliminated but can be reduced by using correct techniques, proper patient positioning, and improved design of CT scanners, or by software provided with the CT scanners.

There are many sources of error in the imaging chain that can result in artifacts. They can generally be classified as artifacts due to the imaging system or artifacts due to the patient. In the following discussion, the geometric alignment, offset and gain correction are caused by imaging system while the scatter and beam hardening artifacts are caused by the nature of object or patient being imaged.

10.9.1 Geometric Misalignment Artifacts

The geometry of a CBCT system is specified using six parameters, namely the three rotation angles (angles corresponding to u, v and w axes in Figure 10.18) and three translations along the principal axis (u, v, w in Figure 10.18). Error in these parameters can result in ring artifact [13],[24], double wall artifact etc., which are visual and hence cannot be misdiagnosed as a pathology. However, very small errors in these parameters can result in blurring of edges and hence misdiagnosis of the size of the pathology, or shading artifacts that could shift the HU number. Hence these parameters must be determined accurately and corrected prior to reconstruction.

10.9.2 Scatter

An incident x-ray photon ejects an electron from the orbit of the atom and consequently a low energy x-ray photon is scattered from the atom. The scattered photon travels at an angle from its incident direction (Figure 10.19). These scattered radiations are detected but are considered as primary radiation. They reduce the contrast of the image and produce blurring. The effect of scatter in the final image is different for conventional radiography and CT. In the case of radiography, the images have poor contrast but in the case of CT, the logarithmic transformation results in a non-linear effect.

Scatter also depends on the type of image acquisition technique.

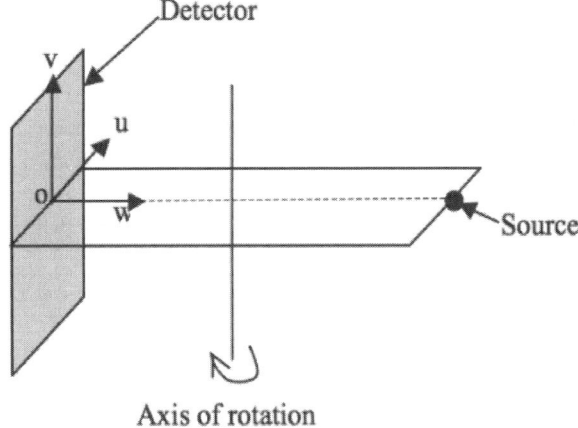

FIGURE 10.18: Parameters defining a cone beam system.

For example, fan-beam CT has less scatter compared to a cone-beam CT due to the smaller height of the beam.

One of the methods to reduce scatter is the air gap technique. In this technique, a large air-gap is maintained between the patient and the detector. Since the scattered radiation at a large angle from the incident direction cannot reach the detector, it will not be used in the formation of the image. It is not always possible to provide an air gap between the patient and the detector, so grids or post-collimators [16],[39] made of lead strips are used to reduce scatter. The grids contain space which corresponds to the photo-detector being detected. The scattered radiation arriving at a large angle will be absorbed by lead and only primary radiations arriving at a small angle from incident direction is detected. The third approach is software correction [53],[68]. Since scatter is a low-frequency structure causing blurring, it can be approximated by a number estimated using beam-stop technique [39]. This, however, does not remove the noise associated with the scatter.

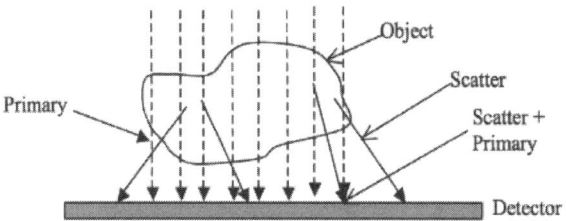

FIGURE 10.19: Scatter radiation.

10.9.3 Offset and Gain Correction

Ideally, the response of a detector must remain constant for a constant x-ray input at any time. But due to temperature fluctuations during acquisition, non-idealities in the production of detectors and variations in the electronic readouts, a non-linear response may be obtained in the detectors. These non-linear responses result in the output of that detector cell being inconsistent with reference to all the neighboring detector pixels. During reconstruction, the non-linear responses produce ring artifacts [39] with their center being located at the isocenter. These circles may not be confused with a human anatomy, as there are no parts which form a perfect circle, but they degrade the quality of the image and hide details and hence need to be corrected. Moreover the detector produces some electronic readout, even when the x-ray source is turned off. This readout is referred to as "dark current" and needs to be removed prior to reconstruction.

Mathematically the flat field and zero offset corrected image (IC) is given by

$$IC(x,y) = \frac{IA - ID}{IF - ID}(x,y) * Average(IF - ID) \tag{10.26}$$

where IA is the acquired image, ID is the dark current image, IF is flat field image, which is acquired at the same technique as the acquired image with no object in the beam. The ratio of the differences is multiplied by the average value of $(IF - ID)$ for gain normalization. This process is repeated for every pixel. The dark field images are to be acquired before each run, as they are sensitive to temperature variations. Other software based correction techniques based on image processing are also used to remove the ring artifacts. They can be classified as pre-processing and post-processing techniques. The pre-processing techniques are based on the fact that the rings in the reconstructed images appear as vertical lines in the sinogram space. Since no feature in an object except those at iso-center can appear as vertical lines, the pixels corresponding to vertical lines can be replaced using estimated pixel values. Even though the process is simple, the noise and complexity of human anatomy present a big challenge in the detection of vertical lines. Another correction scheme is the post-processing technique [39]. The rings in the reconstructed images are identified and removed. Since ring detection is primarily an arc detection technique, it could result in over-correcting the reconstructed image for features that look like arcs. So in supervised ring removal technique, inconsistencies across all views are considered. To determine the position of pixels corresponding to a given ring radius, a mapping that depends on the location of source, object and image is used.

10.9.4 Beam Hardening

The spectrum (Figure 10.2) does not have a unique energy but has a wide range of energies. When such an energy spectrum is incident on a material, the lower energy gets attenuated faster as it is preferentially absorbed than the higher energy. Hence a polychromatic beam becomes harder or richer in higher energy photons as it passes through the material. Since the reconstruction process assumes an "ideal" monochromatic beam, the images acquired using polychromatic beam produce cupping artifacts [1]. The cupping artifact is characterized by a radial increase in intensity from the center of the reconstructed image to its periphery. Unlike ring artifacts, this artifact presents a difficulty, as it can mimic some pathology and hence can lead to misdiagnosis. The cupping artifacts also shift the intensity values and hence present difficulty in quantification of the reconstructed image data. They can be reduced by hardening the beam prior to reaching the patient, using filter made of aluminum, copper etc. Algorithmic approaches for reducing these artifacts have also been proposed.

10.9.5 Metal Artifacts

Metal artifacts are caused by the presence of materials that have a high attenuation coefficient when compared to pathology in the human body. These include surgical clips, biopsy needles, tooth fillings, implants etc. Due to their high attenuation coefficient, metal artifacts produce beam-hardening artifacts (Figure 10.20) and are characterized by streaks emanating from the metal structures. Hence techniques used for removing beam hardening can be used to reduce these artifacts.

In Figure 10.20, the top image is a slice taken at a location without any metal in the beam. The bottom image contains an applicator. The beam hardening causes a streaking artifact that not only renders the metal poorly reconstructed but also adds streaks to nearby pixels and hence makes diagnosis difficult.

Algorithmic approaches [39], [44], [104] to reducing these artifacts have been proposed. A set of initial reconstructions is performed without any metal artifact correction. From the reconstructed image, the location of metal objects is then determined. These objects are then removed from the projection image to obtain synthesized projection. The synthesized projection is then reconstructed to obtain a reconstructed image without metal artifacts.

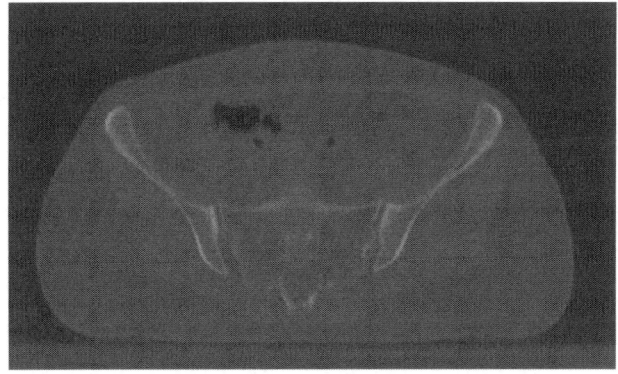

(a) Slice with no metal in the beam

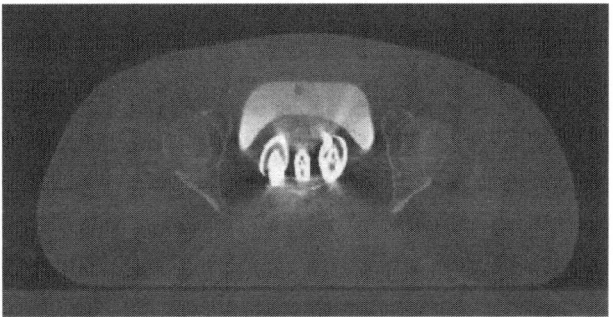

(b) Beam hardening artifact with strong streaks emanating from a metal applicator

FIGURE 10.20: Effect of metal artifact.

10.10 Summary

- A typical x-ray and CT system consists of a x-ray tube, detector and a patient table.

- X-ray is generated by bombarding high-speed electrons on a tungsten target. A spectrum of x-ray is generated. There are two parts to the spectrum: Bremsstrahlung or braking spectrum and the characteristic spectrum.

- The x-ray passes through a material and is attenuated. This is governed by the Lambert Beer law.

- The x-ray after passing through a material is detected using either an ionizing detector or a scintiallation detector such as II or FPD.

- X-ray systems can be either fluoroscopic or angiographic.

- A CT system consists of an x-ray tube and detector, and these are rotated around the patient to acquire multiple images. These images are reconstructed to obtain the slice through a patient.

- The central slice theorem is an analytical technique for reconstructing images. Based on this theorem, it can be proven that the reconstruction process consists of filtering and then backprojection.

- Hounsfield unit is the unit of measure in CT. The unit is a map of the attenuation coefficient of the material.

- CT systems suffer from various artifacts such as misalignment artifact, scatter artifact, beam hardening artifact, and metal artifact.

10.11 Exercises

1. Describe briefly the various parameters that control the quality of x-ray or CT images.

2. An x-ray tube has a acceleration potential of 50kVp. What is the wavelength of the x-ray?

3. Describe the difference in the detection mechanism between II and FPD. Specifically describe the advantages and disadvantages.

4. Allan M. Cormack and Godfrey N. Hounsfield won the 1979 Nobel Prize for creation of CT. Read their Nobel acceptance speech and understand the improvement in contrast and spatial resolution of the images described compared to current clinical images.

5. What is the HU value of a material whose linear attenuation coefficient is half of the linear attenuation coefficient of water?

6. Metal artifact causes significant distortion of an image both in structure and HU value. Using the list of papers in the references, summarize the various methods.

Chapter 11

Magnetic Resonance Imaging

11.1 Introduction

Magnetic Resonance Imaging (MRI) is built on the same physical principle as Nuclear Magnetic Resonance (NMR), which was first described by Dr. Isidor Rabi in 1938 and for which he was awarded the Nobel Prize in Physics in 1944. In 1952, Felix Bloch and Edward Purcell won the Nobel Prize in Physics for demonstrating the use of NMR technique in various materials.

It took a few more decades to apply the NMR principle to imaging the human body. Paul Lauterbur developed the first MRI machine that generated 2D images. Peter Mansfield expanded on Paul Lauterbur's work and developed mathematical techniques that are still part of MRI image creation. For their work, Peter Mansfield and Paul Lauterbur were awarded the Nobel Prize in Physics in 2003.

MRI has developed over the years as one of the most commonly used diagnostic tools by physicians all over the world. It is also popular due to the fact that it does not use ionizing radiation. It is superior to CT for imaging tissue, due to its better tissue contrast.

Unlike the physics and workings of CT, MRI physics is more involved and hence this chapter is arranged differently than the CT chapter. In the x-ray and CT chapter, we began with the construction and generation of x-ray, then discussed the material properties that govern x-ray imaging and finally discussed x-ray detection and image formation. However, in this chapter we begin the discussion with the various laws that govern NMR and MRI. This includes Faraday's law of electromagnetic induction, Larmor frequency and the Bloch equation. This is followed by material properties such as T_1 and T_2 relaxation times and the gyromagnetic ratio and proton density and that govern MRI imaging. This is followed by sections on NMR detection and MRI imaging. With all the physics understood, we discuss the construction of an MRI machine. We conclude with the various modes and potential artifacts in MRI imaging. Interested readers can find more details in [11],[15],[20],[38],[55],[56],[61],[101],[106].

11.2 Laws Governing NMR and MRI

11.2.1 Faraday's Law

Faraday's law is the basic principle behind electric motors and generators. It is also part of today's electric and electric-hybrid cars. It was discovered by Michael Faraday in 1831 and was correctly theorized by James Clerk Maxwell. It states that current is induced in a coil at a rate at which the magnetic flux changes. In Figure 11.1, when the magnet is moved in and out of the coil in the direction shown, a current is induced in the coil in the direction shown. This is useful for electrical power generation, where the flux of the magnetic field is achieved by rotating a powerful magnet inside the coil. The power for the motion is obtained using mechanical means such as potential energy of water (hydroelectric), chemical energy of diesel (diesel engine power plants) etc.

The converse is also true. When a current is passed through a closed circuit coil, it will cause the magnet to move. By constricting the motion of the magnet to rotation, an electric motor can be created. By suitably wiring the coils, an electric generator can thus become an electric motor.

In the former, the magnet is rotated to induce current in the coil while in the latter, a current passed through the coil rotates the magnet.

MRI and NMR use electric coils for excitation and detection. During the excitation phase, the current in the coil will induce a magnetic field that causes the atoms to align in the direction of the magnetic field. During the detection phase, the change in the magnetic field is detected by measuring the induced current.

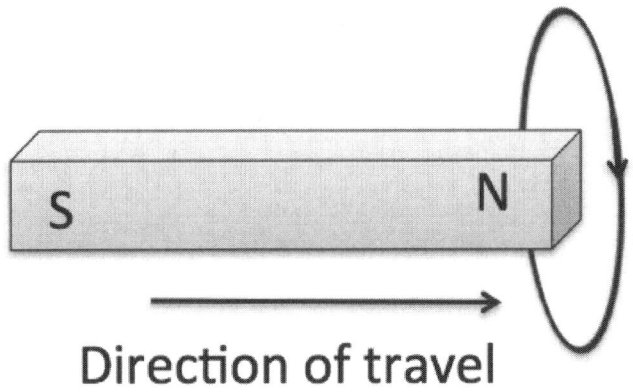

FIGURE 11.1: Illustration of Faraday's law.

11.2.2 Larmor Frequency

An atom (although a quantum level object) can be described as a spinning top. Such a top will be precessing about its axis at an angle as shown in Figure 11.2. The frequency of the precession is an important factor and is described by the Larmor equation (Equation 11.1).

$$f = \gamma B \qquad (11.1)$$

where γ is the gyromagnetic ratio, f is the Larmor frequency, and B is the strength of the external magnetic field.

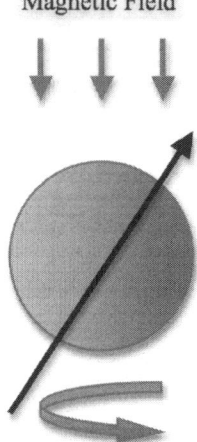

FIGURE 11.2: Precessing of nucleus in a magnetic field.

11.2.3 Bloch Equation

An atom in a magnetic field is aligned in the direction of the field. An RF pulse (to be introduced later) can be applied to change the orientation of the atom. If the magnetic field is pointing in the z-direction, the atom will be aligned in the z-direction. If a pulse of sufficient strength is applied, the atom can be oriented in x- or y-direction or sometimes even in the z direction, the direction opposite to its original.

If the RF pulse is removed, the atom returns to its original z-direction orientation. During the process of moving from the xy-direction to the z-direction, the atom traces a spiral motion, described by the Bloch equations (Equation 11.2).

$$\begin{aligned} M_x &= e^{-\frac{t}{T_2}} \cos \omega t \\ M_y &= e^{-\frac{t}{T_2}} \sin \omega t \\ M_z &= M_0 (1 - e^{-\frac{t}{T_1}}) \end{aligned} \quad (11.2)$$

The equations can be easily visualized by plotting them in 3D (Figure 11.3). At time $t = 0$, the value of M_z is zero. This is due to the fact the atoms are oriented in the xy-plane and hence their net magnetization is also in the xy-plane and not in the z-direction. When the RF pulse is removed, the atoms begin to orient in the z-direction (their original direction before RF pulse). This change along the xy-plane is an exponential decay in amplitude change and sinusoidal in directional change. Thus, the net magnetization reduces over time exponentially while sinusoidally changing direction in the xy-plane. At $t = infinity$, the M_x and M_y reach 0 while M_z reaches the original value of M_0.

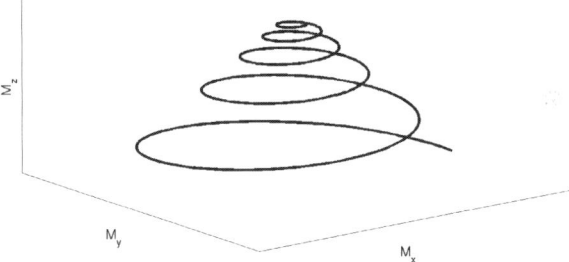

FIGURE 11.3: Bloch equation as a 3D plot.

11.3 Material Properties

11.3.1 Gyromagnetic Ratio

The gyromagnetic ratio of a particle is the ratio of its magnetic dipole moment to its angular momentum. It is a constant for a given nuclei. The values of gyromagnetic ratio for various nuclei are given in Table 11.1. When an object containing multiple materials (and hence different nuclei) is placed in a magnetic field of certain strength, the precessional frequency is directly proportional to the gyromagnetic ratios based on the Larmor equation. Hence, if we measure the precessional

Nuclei	γ (MHz/T)
H^1	42.58
P^{31}	17.25
Na^{23}	11.27
C^{13}	10.71

TABLE 11.1: An abbreviated list of the nuclei of interest to NMR and MRI imaging and their gyromagnetic ratios.

frequency, we can distinguish the various materials. For example, the gyromagnetic ratio is 42.58 MHz/T for a hydrogen nucleus while it is 10.71 MHz/T for a carbon nucleus. For a typical clinical MRI machine, a common magnetic field strength (B) is 1.5T. Hence the precessional frequency of hydrogen atom is 63.87 MHz and that of the carbon is 16.07 MHz.

11.3.2 Proton Density

The second material property that is imaged is the proton density or spin density. It is the number of "mobile" hydrogen nuclei in a given volume of the sample. The higher the proton density, the larger the response of the sample in NMR or MRI imaging.

The response to NMR and MRI is not only dependent on the density of hydrogen nucleus but also its configuration. A hydrogen nucleus connected to oxygen responds differently compared to the one connected to carbon atom. Also, a tightly bound hydrogen atom does not produce any noticeable signal. The signal is generally produced by an unbound or free hydrogen nucleus. Thus, the hydrogen atom in tissue that is loosely bound produces a stronger signal. Bone on the other hand has hydrogen atoms that are strongly bound and hence produces a weaker signal.

Table 11.2 lists the proton density of common materials. It can be seen from the table that the proton density for bone is low compared to white matter. Thus, the bone responds poorly to the MRI signal. One exception in the table is the response of fat to the MRI signal.

Magnetic Resonance Imaging

Biological material	Proton or spin density
Fat	98
Grey matter	94
White matter	100
Bone	1-10
Air	< 1

TABLE 11.2: List of biological materials and their proton or spin density.

Although fat consists of a large number of protons, it responds poorly to the MRI signal. This is due to the long chain of molecules found in fat that immobilize hydrogen atom.

11.3.3 T_1 and T_2 Relaxation Times

There are two relaxation times that characterize the various regions in an object and can help distinguish them in an MRI image. They characterize the response of an atom in the Bloch equation.

Consider the image shown in Figure 11.4. A strong magnetic field B_0 is applied in the direction of the z-axis. This causes a net magnetization of M_0 in the z-axis to increase from zero. The increase is initially rapid but then slows down. It is given by Equation 11.3 and graphically represented by Figure 11.5.

$$M_z = M_0(1 - e^{-\frac{t}{T_1}}) \tag{11.3}$$

The time for the net magnetization to reach a value within e (i.e. $M_0 - M_z = \frac{M_0}{e}$) is called the T_1 relaxation time. Since T_1 deals with both magnetization and demagnetization along the longitudinal direction (z-axis), T_1 is also referred to as **longitudinal relaxation time**.

During an MRI image acquisition, in addition to the external magnetic field, an RF pulse is applied. This RF pulse disturbs the equilibrium and reduces M_z. The protons are not in isolation from other atoms but instead are bound tightly by a lattice. When the RF pulse is

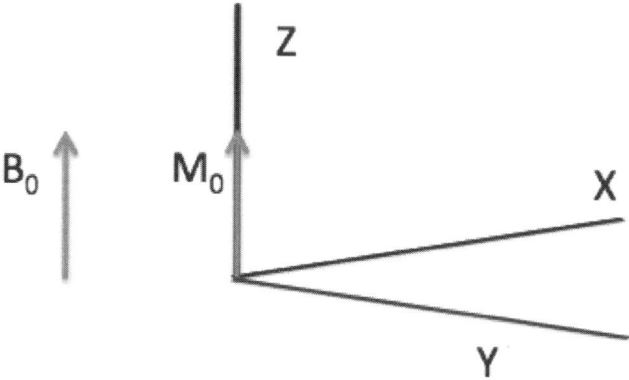

FIGURE 11.4: T_1 magnetization.

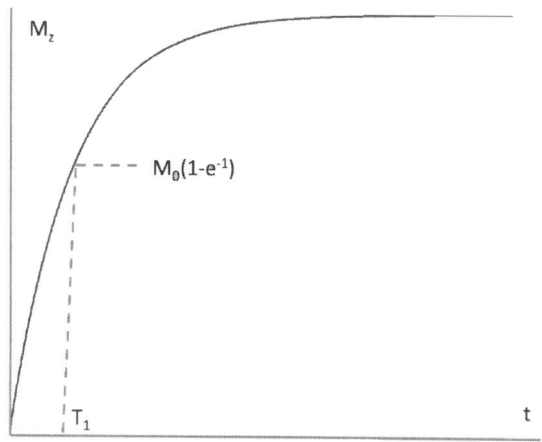

FIGURE 11.5: Plot of T_1 magnetization.

removed, the protons return to equilibrium, which causes a decrease in M_{xy} or transverse magnetization. This is accomplished by transferring energy to other atoms and molecules in the lattice. The time constant for the magnetization decay in the xy-axis is called T_2 or **spin-lattice relaxation time**. It is governed by Equation 11.4 and is graphically represented by Figure 11.6.

$$M_{xy} = M_{xy0} e^{-\frac{t}{T_2}} \tag{11.4}$$

Biological material	T_1 (ms)	T_2 (ms)
Cerebrospinal fluid	2160	160
Grey matter	810	100
White matter	680	90
Fat	240	80

TABLE 11.3: List of biological materials and their T_1 and T_2 values for field strength of 1.0 T.

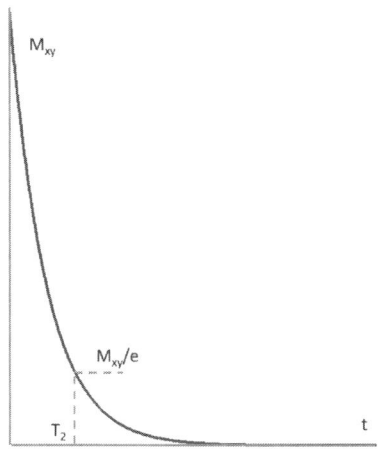

FIGURE 11.6: Plot of T_2 de-magnetization.

T_1 and T_2 are independent of each other but T_2 is generally smaller than or equal to T_1. This will be evident from the Table 11.3, which lists T_1 and T_2 values for some common biological materials. The value of T_1 and T_2 are dependent on the strength of the external magnetic field (1.0T in this case).

11.4 NMR Signal Detection

As discussed earlier, the presence of a strong magnetic field aligns the proton in the object in the direction of the magnetic field. The most interesting phenomenon happens when an RF pulse is applied to the object in the presence of the main magnetic field.

Due to the strong magnetic field (B_0), the protons align themselves with it. They also precess at the Larmor frequency. This is the equilibrium state of the proton under the magnetic field. When an RF pulse is applied using the transmitting coil to the cartoon head (Figure 11.7), the proton orientation changes and in some cases it flips in the negative direction while precessing at the Larmor frequency. Due to the flip, the net magnetization is in the direction opposite to the direction of the main magnetic field. When the RF pulse is removed, the protons flip back to the positive direction and hence reach their equilibrium state. During this process, an electric current is induced in the receiving coil due to changing magnetic field. This based on Faraday's law which was discussed previously. The signal obtained in the receiving coil is shown in Figure 11.8. The signal reduces in its intensity over time due to free induction decay (FID) and the time for the protons to reach their equilibrium state, or the "relaxed" state, is called the "relaxation time".

The signal is a plot over time. This signal contains details of the frequencies of various protons in the object. The frequency distribution can be obtained by using Fourier transform.

11.5 MRI Signal Detection or MRI Imaging

In this section, we will learn methods for obtaining images using MRI. The actual imaging process begins with selection of a section of the object being imaged and placing that section under a magnetic field in a process called slice selection. An MRI signal can only be achieved by changing the orientation of the proton under the magnetic field. This is achieved by applying RF pulse in the other two orthogonal directions during phase and frequency encoding. All these activities need to be

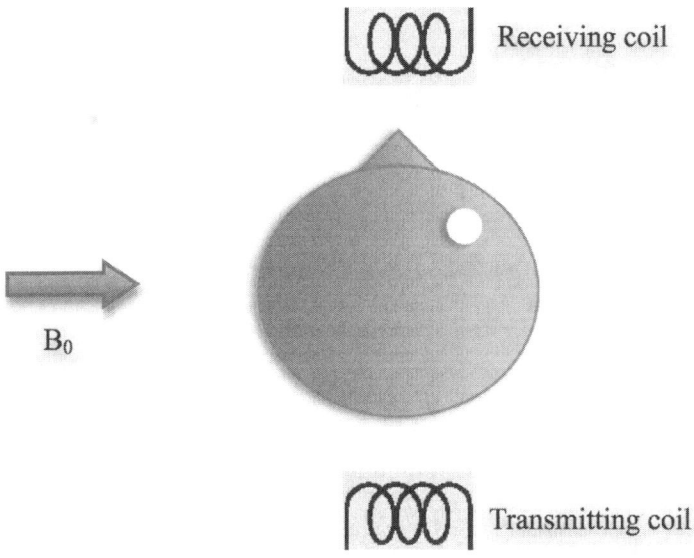

FIGURE 11.7: Net magnetization and effect of RF pulse.

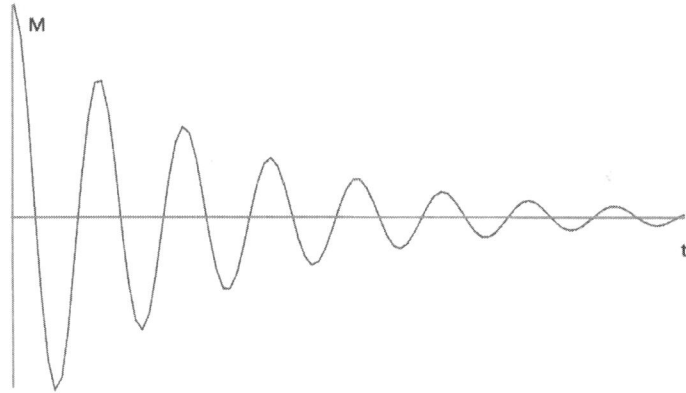

FIGURE 11.8: Free induction decay.

timed so that an MRI image can be obtained. This timing process is called pulse sequence. We will discuss each of these in detail in the subsequent sections.

11.5.1 Slice Selection

Slice selection is achieved by applying the magnetic field on an object along one of the orthogonal directions in generally the z-direction or axial direction. Application of the magnetic field causes the protons in that section to orient themselves in the direction of the magnetic field and limits the imaging to this section. The slices that are not under the magnetic field are oriented randomly and hence will not be affected by the subsequent application of the magnetic field or RF pulses.

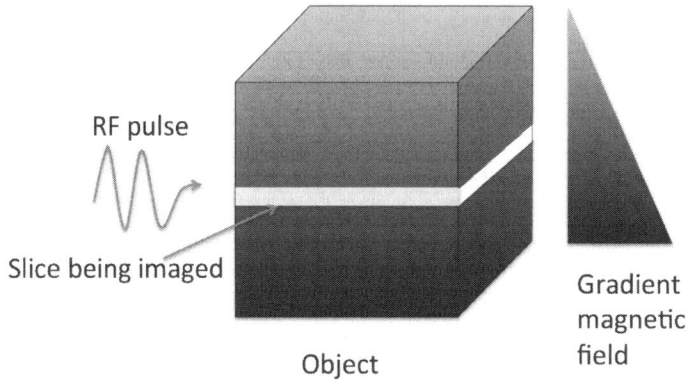

FIGURE 11.9: Slice selection gradient.

11.5.2 Phase Encoding

Phase encoding gradient is generally applied in the x-, y- or z-direction. Due to the application of slice selection gradient, the various protons are oriented in the z-direction (say). They will be spinning in phase with each other. By applying a gradient along the y-direction (say), the protons along a given y-location will spin with the same phase and the other y-locations will spin out of phase. Since every y-location can be identified using its phase, it can be concluded that the protons are encoded with reference to phase. The same argument can be extended if the phase encoding gradient is in the x-direction. If the MRI image has N pixel locations, the phase encoding gradient is chosen such that the phase shift between adjacent pixel is given by Equation 11.5. This ensures that two coordinates do not share the same phase.

$$\phi = \frac{360}{\text{Number of pixels along } x \text{ or } y} \qquad (11.5)$$

11.5.3 Frequency Encoding

Frequency encoding gradient is applied in the x-, y- or z- direction. After the application of phase encoding gradient along the y-direction, all the protons along a given y-location will be precessing at the same phase. When a frequency encoding gradient is applied along the x-direction, protons for a given x-location will receive the same magnetic field. Hence these protons will precess at the same frequency. **Thus, with the application of both phase and frequency encoding gradient, every x-, y- point in the object will have a unique phase and frequency.**

11.6 MRI Construction

A simple model (Figures 11.10 and 11.11) of an MRI will consist of:

- Main magnet
- Gradient magnet
- Radio-frequency coils
- Computer for processing the signal

11.6.1 Main Magnet

The main magnet generates a strong magnetic field. A typical MRI machine used for medical diagnosis is around 1.5T, which is 30,000 times stronger than the earth's magnetic field.

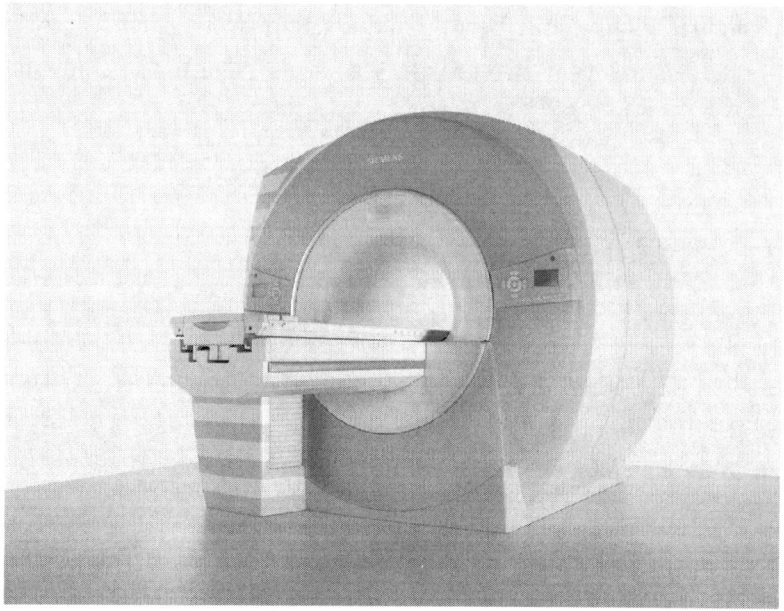

FIGURE 11.10: Closed magnet MRI machine. Original image reprinted with permission from Siemens AG.

The magnets could be permanent magnet, electromagnet or superconducting magnets. An important criterion for choosing a magnet is its ability to produce uniform magnetic field. Permanent magnets are cheaper but the magnetic field is not uniform. Electromagnets can be manufactured to close tolerance, so that the magnetic field is uniform. They generate a lot of heat, which limits the magnetic field strength. Superconducting magnets are electromagnets that are cooled by superconducting fluids such as liquid nitrogen or helium. These magnets have a homogeneous magnetic field and high field strength but they are expensive to operate.

11.6.2 Gradient Magnet

As described earlier, a uniform magnetic field cannot localize the various parts of the object. Hence gradient magnetic fields are used. Based on Faraday's law, a magnetic field can be generated by the ap-

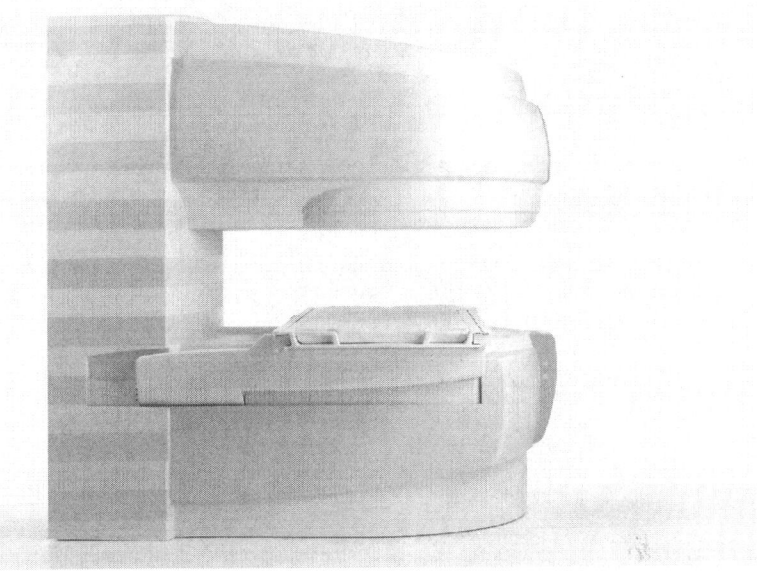

FIGURE 11.11: Open magnet MRI machine. Original image reprinted with permission from Siemens AG.

plication of current to a coil, also known as gradient coils. Since gradient needs to be generated in all three directions, gradient coils are configured to generate fields in all three directions.

11.6.3 RF Coils

An RF coil is composed of loops of conducting materials such as copper. It generates a magnetic field with the passage of current. This process is called transmitting signal. Similarly, a rapidly changing magnetic field generates current in the coil which can be measured. This is accomplished using a receiving coil. In some cases, the same coil can transmit and receive signals. Such coils are called transceivers. An example of a brain imaging coil is shown in Figure 11.12. Specialized coils are created for different parts being imaged.

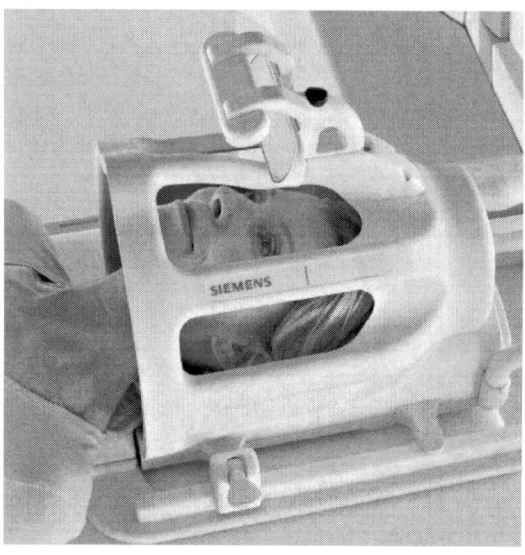

FIGURE 11.12: Head coil. Original image reprinted with permission from Siemens AG.

11.6.4 K-Space Imaging

In Section 11.4, we discussed that the protons regain their orientation after the removal of the RF pulse. During this process, an FID signal (Figure 11.8) is induced in the coil. The FID signal is a plot over time of the change in the net-magnetization in the transverse plane. This signal contains various frequencies that can be obtained using Fourier transformation (Chapter 6). This signal is a 1D signal, as the originating signal is also 1D.

In Section 11.5, we also discussed that the three magnetic field gradients allow localization of signal. The three magnetic fields are applied, and the signal obtained for each condition is readout. This 1D signal fills one horizontal line in the frequency space (Figure 11.13). By repeating the signal generation process for all conditions, the various horizontal lines can be filled.

It can be proven that the image acquired in Figure 11.13 is the Fourier transform of the MRI image. A simple inverse Fourier transform can be used to obtain the MRI image (Figure 11.14). Figure 11.14(a)

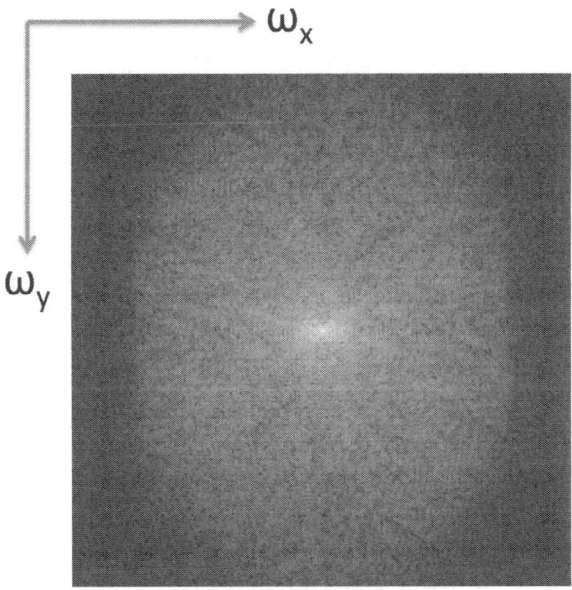

FIGURE 11.13: k-space image.

is the image obtained by filling the k-space and Figure 11.14(b) is obtained using inverse Fourier transform of the first image.

11.7 T_1, T_2 and Proton Density Image

A typical MRI image consists of T_1, T_2 and proton density weighted components. It is possible to obtain pure T_1, T_2 and proton density weighted image but it is generally time consuming. Such images are used for discussion to emphasize the role each of these components play in MRI imaging.

Figure 11.15(a) is a T_1 weighted image (i.e. the pixel values are dependent on the T_1 relaxation time). Similarly, Figure 11.15(b) and Figure 11.15(c) are T_2 and proton density weighted images respectively.

Bright pixels in a T_1 weighted image correspond to fat, while the

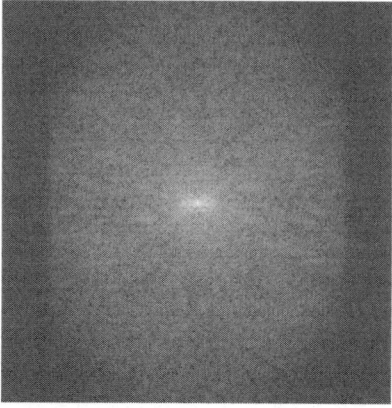

(a) Image obtained by filling the k-space.

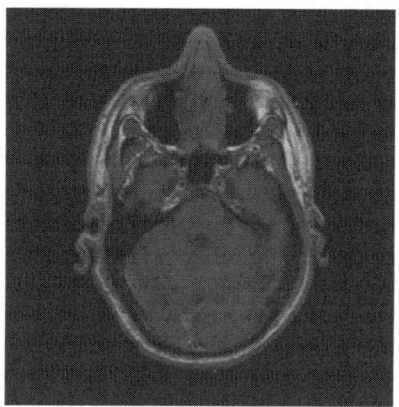

(b) Inverse Fourier transform of k-space image.

FIGURE 11.14: k-space reconstruction of MRI images.

same pixels appear darker in a T_2 weighted image and vice versa. A proton density image is useful in identifying the pathology of the object.

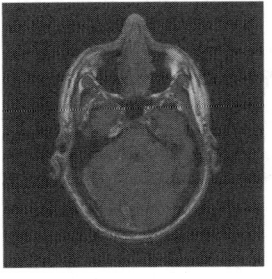 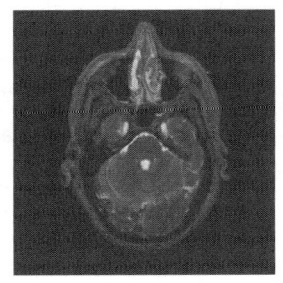 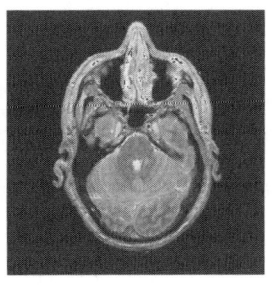

(a) T_1 weighted image.

(b) T_2 weighted image.

(c) Proton density weighted image.

FIGURE 11.15: T_1, T_2 and proton density image. Courtesy of the Visible Human Project.

11.8 MRI Modes or Pulse Sequence

So far, we have learned about the various controls such as gradient magnitude along the three axes and the RF pulse that tilts the orientation of the protons. In this section, we will combine these four controls to produce images that are medically and scientifically useful. This process consists of performing different operations at different times and is generally shown using a pulse sequence diagram. In this diagram, each control receives its own row of operation. The time progresses to the right of each row. Some of these pulse sequences are discussed below. In each case, a certain set of operations or sequences is repeated at regular intervals called repetition time or TR. TE is defined as the time between the start of the first RF pulse and the time to reach peak of echo (or output signal).

11.8.1 Spin Echo Imaging

Spin echo pulse sequence (Figure 11.16) is one of the simplest and most commonly used pulse sequences. It consists of a 90^0 pulse followed by 180^0 pulse at TE/2. During both the pulses, the gradient magnitude along the z-axis is kept on. An echo is produced at time TE while the gradient along the x-axis is kept on so that localization information can be obtained. This process is repeated after every time to repeat (TR). The last 90^0 pulse in the figure is the start of the next sequence.

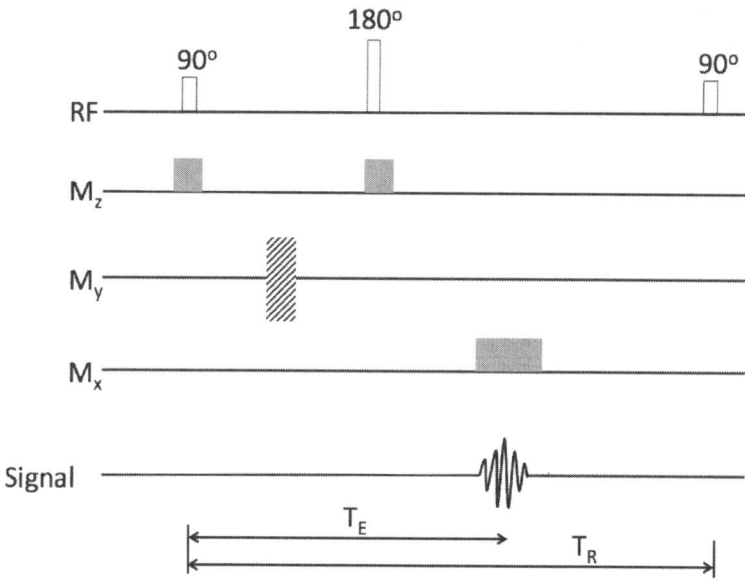

FIGURE 11.16: Spin echo pulse sequence.

11.8.2 Inversion Recovery

The inversion recovery pulse sequence (Figure 11.17) is similar to the spin echo sequence except for a 180^0 pulse applied before the 90^0 pulse.

The 180^0 pulse causes the net-magnetization vector to be inverted along the z-axis. Since the inversion cannot be measured in planes other than xy-plane, a 90^0 pulse is applied. The time between the two pulses is called the inversion time or TI. The gradient magnetic field along the z-axis is kept on during both the pulses. The gradient is applied while reading the echo, so that the localization information can be obtained.

This process is repeated at regular intervals of TR. The last 180^0 pulse in the figure is the start of the next sequence.

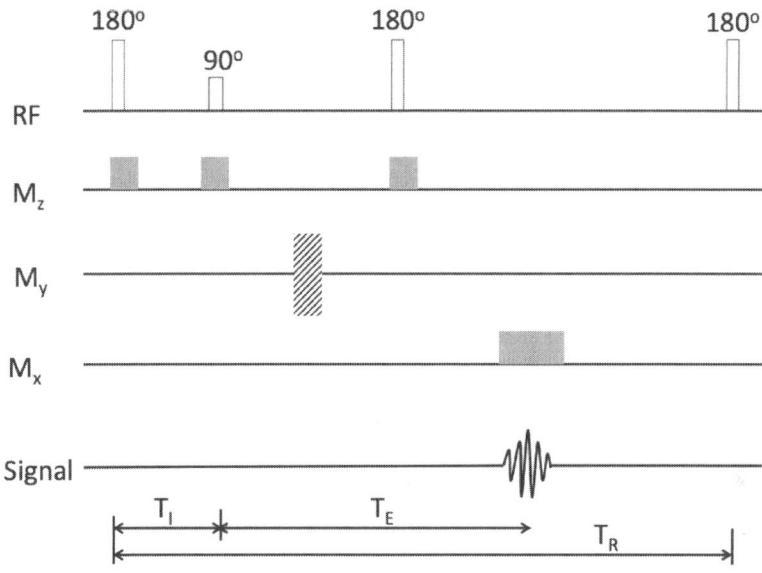

FIGURE 11.17: Inversion recovery pulse sequence.

11.8.3 Gradient Echo Imaging

Gradient echo imaging pulse sequence (Figure 11.18) consists of only one pulse of 90^0 and is one of the simplest pulse sequences. The flip angle could be any angle and 90^0 was chosen as one example. The slice selection gradient is kept on during the application of the 90^0 pulse. At the end of the pulse, a gradient magnetic field is applied along the y-axis. A negative gradient is applied along the x-axis at the same time. The x-axis gradient is then switched to positive gradient while the echo is read. Since there are fewer pulses and the gradient are switched on at consecutive intervals, it is one of the fastest imaging techniques. The last 90^0 pulse in the figure is the start of the next sequence.

By using various settings for TR and TE, it is possible to obtain

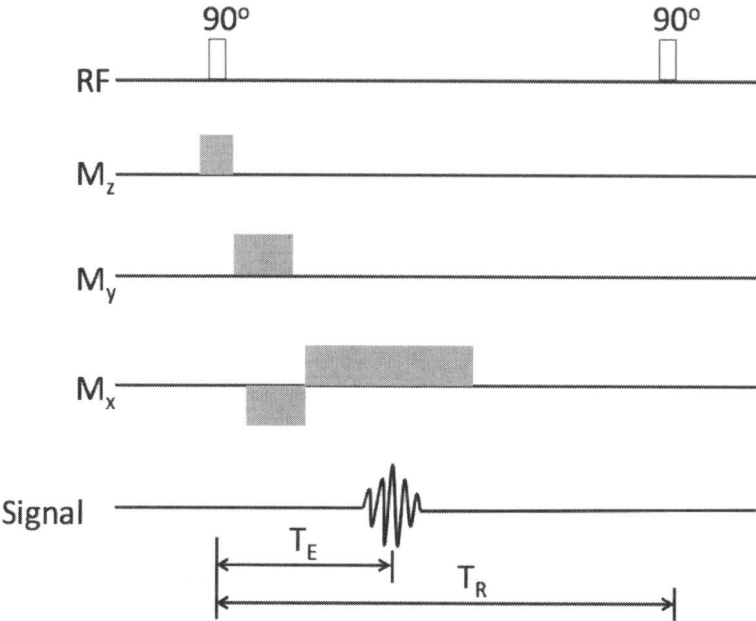

FIGURE 11.18: Gradient echo pulse sequence.

Weighted image	TR	TE
T_1	Short	Short
T_2	Long	Long
Proton density	Long	Short

TABLE 11.4: TR and TE settings for various weighted images.

images that are weighted for T_1, T_2 and proton density. A list of such parameters is shown in Table 11.4.

11.9 MRI Artifacts

Image formation in MRI is complex with interaction of various parameters such as homogeneity of the magnetic field, homogeneity of the applied RF signal, shielding of the MRI machine, presence of metal that can alter the magnetic field etc. Any deviation from ideal conditions will result in an artifact that could either change the shape or the pixel intensity in the image. Some of these artifacts are easily identifiable. For example, a metal artifact leaves identifiable streaks or distortions. A few other artifacts are not easily identifiable. An example of such an artifact that is not easily identifiable is the partial volume artifact.

These artifacts can be removed by creating close to ideal conditions. For example, to ensure that there are no metal artifacts, it is important that the patient does not have any implanted metal objects. Alternately, a different imaging modality such as CT or a modified form of MRI imaging can be used.

Artifacts are generally classified into two categories: patient related and machine related. The motion artifact and metal artifact are patient related while inhomogeneity and partial volume artifacts are machine related. An image may contain artifacts from both categories. There are many other artifacts. Interested readers must check the references in Section 11.1.

11.9.1 Motion Artifact

A motion artifact can be due to either motion of the patient or the motion of the organs in a patient. The organ motions occur due to cardiac cycle, blood flow, breathing. In MRI facilities with poor shielding or design, moving large ferromagnetic objects such as automobiles, elevators etc., can cause inhomogeneity in the magnetic field that in turn can cause motion artifacts.

Motion due to cardiac cycle can be controlled by gating, a process of timing the image acquisition with the heart cycle. In some cases, simple breath holding can be used to compensate for the motion artifact.

Figure 11.19 is an example of a slice reconstructed with and without motion artifact. The motion artifact in Figure 11.19(b) has resulted in significant degradation of the image quality, making clinical diagnosis difficult.

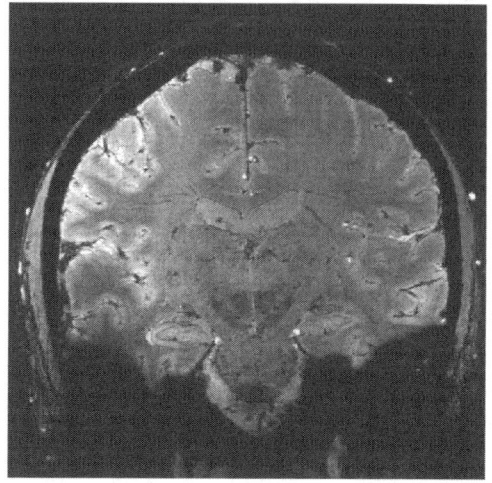

(a) Slice with no motion artifact

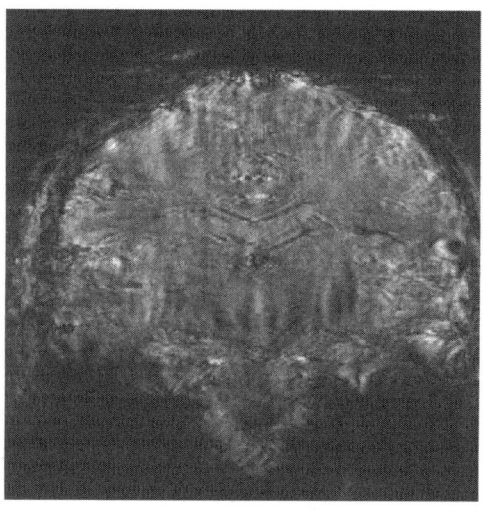

(b) Slice with motion artifact

FIGURE 11.19: Effect of motion artifact on MRI reconstruction. Original images reprinted with permission from Dr. Essa Yacoub, University of Minnesota.

11.9.2 Metal Artifact

Ferromagnetic materials such as iron strongly affect the magnetic field, causing inhomogeneity. In Figure 11.20, the arrows in the two images indicate the direction of the magnetic field. In the left image, the magnetic field surrounds a non-metallic object such as tissue. The presence of the tissue does not change the homogeneity of the magnetic field. In the right image, a metal object is placed in the magnetic field. The magnetic field is distorted close to the object.

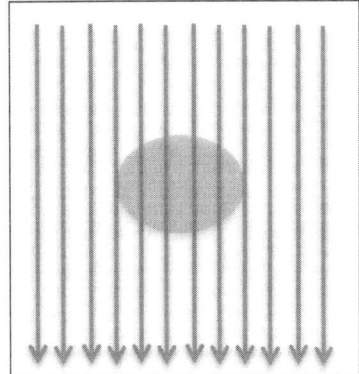 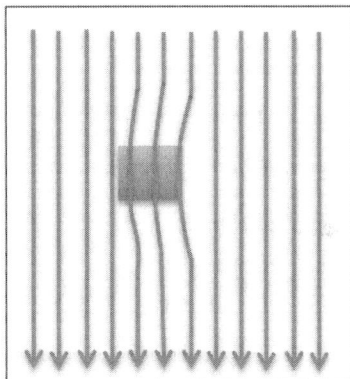

FIGURE 11.20: Metal artifact formation.

The reconstruction process assumes that the field is homogeneous. Thus, it assumes that all points with the same magnetic field strength will have the same Larmor frequency. This variation from ideality causes metal artifact.

The effect is more profound in the case of ferromagnetic materials such as iron, stainless steel etc. It is less profound in metals such as titanium and other alloys. If MRI is the preferred modality for imaging a patient with metal implants, a low field-strength magnet can be used.

11.9.3 Inhomogeneity Artifact

This artifact is similar in principle to the metal artifact. In the case of metal artifact, the inhomogencity is caused by the presence of metallic objects. In the case of the inhomogeneity artifact, the magnetic field is not uniform due to a defect in the design or manufacture of the magnet.

The artifact can occur due to both the main magnetic field (B_0) or due to gradient magnetic field. In some cases, the main magnetic field may not be uniform across the patient and will change from center to periphery.

The artifact results in distortion depending on the variation of magnetic field across the patient. If the variations are minimal, the shading artifact results.

11.9.4 Partial Volume Artifact

This artifact is caused by imaging using large voxel size, causing two nearby object intensities or pixel intensities to be averaged. This artifact generally affects long and thin objects as their intensity changes rapidly in the direction perpendicular to their long axis.

The artifact can be reduced by increasing the spatial resolution, which results in increased number of voxels in the image and consequently longer acquisition time.

11.10 Summary

- MRI is an non-radiative high-resolution imaging technique.

- It works on Faraday's law, Larmor frequency and Bloch equation.

- It is based on physical principles such as T_1 and T_2 relaxation time, proton density and gyromagnetic ratio.

- Atoms in a magnetic field are aligned in the direction of the magnetic field. An RF pulse can be applied to change their orientation. When the RF pulse is removed, the atoms reorient and the current generated by this process can be measured. This is the basic principle of NMR.

- In MRI, the basic principle of NMR is used along with slice selection, phase encoding and frequency encoding gradient to localize the atoms.

- An MRI machine consists of a main magnet, gradient magnets, an RF coil and a computer for processing.

- The various parameters that control MRI image acquisition are diagrammatically represented as a pulse sequence diagram.

- MRI suffers from various artifacts. These artifacts can be classified as either patient related or machine related.

11.11 Exercises

1. Calculate the Larmor frequency for all atoms listed in Table 11.1.

2. Explain the plot in Figure 11.3 using the Equation 11.2.

3. If the plot in Figure 11.3 is viewed looking down in the z direction, the magnetic field path will appear as a circle. Why?

    Solution: The values of M_x and M_y have cos and sin dependencies, similar to the parametric form of a circle.

4. Explain why T_2 is generally smaller or equal to T_1.

5. Before k-space imaging was used, image reconstruction was achieved using a back-projection technique similar to CT. Write a report about this technique.

6. We discussed a few of the artifacts seen in MRI images. Identify two more artifacts and list their causes, symptoms and method to overcome these artifacts.

7. MRI is generally safe compared to CT. Yet it is important to take precautions during MRI imaging. List some of these precautions.

Chapter 12

Light Microscopes

12.1 Introduction

The modern light microscope was created in the 17th century, but the origin of its important component, the lens, dates back more than three thousand years. The ancient Greeks used lenses as burning glasses, by focusing the sun's rays. In later years, lenses were used to create glasses in Europe in order to correct for vision problems. The scientific use of lenses can be dated back to the 16th century with the creation of compound microscopes. Robert Hooke, an English physicist, was the first person to describe cells using a microscope. Antonie van Leeuwenhoek, a Dutch physicist, improved on the lens design and made many important discoveries. For all his research efforts, he is referred to as "the father of microscopy".

We begin this chapter with an introduction to the various physical principles that govern image formation in light microscopy. These include geometric optics, diffraction limit of the resolution, the objective lens, and finally numerical aperture. The aim of a microscope is to magnify an object while maintaining good resolving power (i.e., the ability to distinguish two objects that are nearby). Geometric optics is the magnification achieved by a microscope. The diffraction limit, the objective lens, and the numerical aperture determine the resolving power of a microscope. We apply these principles during the discussion on design of a simple wide-field microscope. This is followed by the fluorescence microscope that not only images the structure but also encodes the functions of the various parts of the specimen. We then discuss confocal and Nipkow disk microscopes that offer better contrast resolution compared to wide-field microscopes. We conclude with a discussion on choosing a wide-field or confocal microscope for a given imaging task. Interested readers can refer to [4], [18],[35],[62],[81],[101] for more details.

12.2 Physical Principles

12.2.1 Geometric Optics

A simple light microscope of today is shown in Figure 12.1. It consists of an eyepiece, an objective lens, the specimen to be viewed and a light source. As the name indicates, the eyepiece is the lens for viewing the sample. The objective is the lens closest to the sample. The eyepiece and the objective lens are typically compound convex lenses. With the introduction of digital technology, the viewer does not necessarily look at the sample through the eyepiece but instead a camera acquires and stores the image.

The lenses used in a microscope have magnification. The magnification for the objective can be defined as the ratio of the height of the image formed to the height of the object. Applying triangular inequality (Figure 12.2), we can also obtain the magnification, m as the ratio of d_1 to d_0.

$$m = \frac{h_1}{h_0} = \frac{d_1}{d_0} \tag{12.1}$$

A similar magnification factor can be obtained for the eyepiece as well. The total magnification of the microscope can be obtained as the product of the two magnifications.

$$M = m_{\text{objective}} * m_{\text{eyepiece}} \tag{12.2}$$

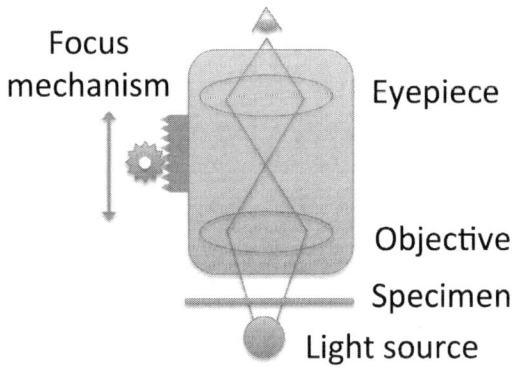

(a) A schematic of the light microscope

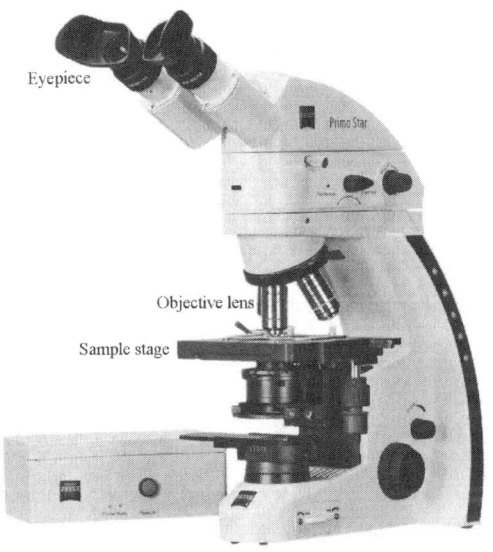

(b) The light microscope. Original image reprinted with permission from Carl Zeiss Microscopy, LLC.

FIGURE 12.1: Light microscope.

12.2.2 Numerical Aperture

Numerical aperture defines both the resolution and the photon-collecting capacity of a lens. It is defined as:

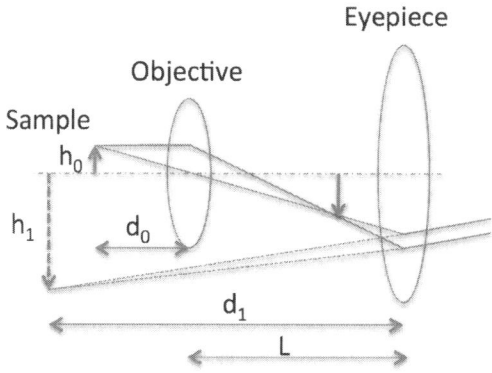

FIGURE 12.2: Schematic of the light microscope.

$$NA = n \sin \theta \qquad (12.3)$$

where θ is the angular aperture or the acceptance angle of the aperture and n is the refractive index.

For high-resolution imaging, it is critical (as will be discussed later) to use an objective with a high numerical aperture. Figure 12.3 is a photograph of an objective with all the parameters embossed. In this example, 20X is the magnification and 0.40 is the numerical aperture.

12.2.3 Diffraction Limit

Resolution is an important characteristic of an imaging system. It defines the smallest detail that can be resolved (or viewed) using an optical system like the microscope. The limiting resolution is called the diffraction limit. We know that electromagnetic radiations have both particle and wave natures. The diffraction limit is due to the wave nature. The Huygens-Fresnel principle suggests that an aperture such as a lens creates secondary wave sources from an incident plane wave. These secondary sources create an interference pattern and produce the Airy disk.

Based on the diffraction principles, we can derive the resolving

FIGURE 12.3: Markings on the objective lens. Original image reprinted with permission from Carl Zeiss Microscopy, LLC.

power of a lens. It is the minimum distance between two adjacent points that can be distinguished through a lens. It is defined as:

$$RP = \frac{0.61\lambda}{NA} \qquad (12.4)$$

If a microscope system consists of both objective and eyepiece, then the formula has to be modified to:

$$RP = \frac{1.22\lambda}{(NA_{obj} + NA_{eye})} \qquad (12.5)$$

where NA_{obj} and NA_{eye} are the numerical apertures of the objective and eyepiece respectively.

The aim of any optical imaging system is to improve the resolving power or reduce the value of RP. This can be achieved by decreasing the wavelength, increasing the aperture angle or increasing the refractive index. Since this discussion is on the optical microscope, we are limited to the visible wavelength of light. X-rays, gamma rays etc. have shorter wavelengths compared to visible light and hence better resolving power. The refractive index (discussed later) of air is 1.00. The refractive index of mediums used in microscopy imaging is generally greater than 1.00 and hence improves resolving power.

Two points separated by large distances will have distinct Airy disks and hence can be easily identified by an observer. If the points are close (middle image in Figure 12.4), the two Airy disks begin to overlap. If the distance between points is further reduced (left image in Figure 12.4), they begin to further overlap. The two peaks approach and the limit at which the human eye cannot separate the two points is referred to as the Rayleigh Criterion.

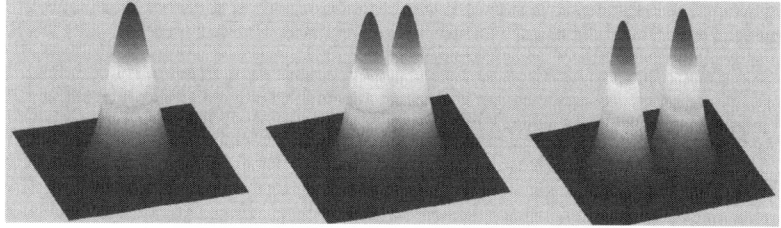

FIGURE 12.4: Rayleigh Criterion.

12.2.4 Objective Lens

In the setup shown in Figure 12.1, the two sources of magnification are the objective lens and eyepiece. Since the objective is the closest to the specimen, it is the largest contributor of magnification. Thus, it is critical to understand the inner workings of the objective lens and also the various choices.

We begin the discussion with the refractive index. It is a dimensionless number that describes how electromagnetic radiation passes through various mediums. The refractive index can be seen in various phenomena such as rainbows, separation of visible light by prisms etc. The refractive index of the lens is different from that of the spec-

Medium	Refractive Index
Air	1.0
Water	1.3
Glycerol	1.44
Immersion oil	1.52

TABLE 12.1: List of the commonly used media and their refractive indexes.

imen. This difference in refractive index causes the deflection of light. The refractive index between the objective lens and the specimen can be matched by submerging the specimen in a fluid (generally called medium) with the refractive index close to the lens.

Table 12.1 shows commonly used media and their refractive indexes. Failure to match the refractive index will result in loss of signal, contrast and resolving power.

To summarize, the objective lens selection is based on the following parameters:

1. Refractive index of the medium

2. Magnification needed

3. Resolution which in turn is determined by the choice of numerical aperture

12.2.5 Point Spread Function (PSF)

Point Spread Function (PSF) refers to the response of an optical system to a point input or point object as consequence of diffraction. A point-like object will appear larger in the image. When a point source of light is passed through a pinhole aperture, the resultant image on a focal plane is not a point but instead the intensity is spread to multiple neighboring pixels. In other words, the point image is blurred by the PSF.

PSF is dependent on the numerical aperture of the lens. A lens with a high numerical aperture produces PSF of smaller width.

12.2.6 Wide-Field Microscopes

Light microscopes can be classified into different types depending on the method used to generate contrast and acquire images, illuminate samples etc. The microscope that we have described is called a wide-field microscope. It suffers from poor spatial resolution (without any computer processing), and poor contrast resolution due to the effect of PSF discussed in the previous section.

12.3 Construction of a Wide-Field Microscope

A light microscope (Figure 12.1) is designed to magnify the image of a sample using multiple lenses. It consists of the following:

1. Eyepiece

2. Objective

3. Light source

4. Condenser lens

5. Specimen stage

6. Focus knobs

The eyepiece is the lens closest to the eye. Modern versions of the eyepiece are compound lenses in order to compensate for aberrations. It is interchangeable and can be replaced by eyepieces of different magnification depending on the nature of object being imaged.

The objective is the lens closest to the object. These are generally compound lenses in order to compensate for aberrations. They are characterized by three parameters: magnification, numerical aperture and the refractive index of the immersion medium. The objectives are interchangeable and hence modern microscopes also contain a turret that contains multiple objectives to enable easier and faster switching between different lenses. The objective might be immersed in oil to match the refractive index and increase the numerical aperture and consequently increase the resolving power.

The light source is at the bottom of the microscope. It can be tuned to adjust the brightness in the image. If the lighting is poor the contrast of the resultant image will be poor while excess light might saturate the camera recording the image. Excess light may cause photo bleaching. The most commonly used illumination method is the Köhler illumination, designed by August Köhler in 1893. The previous methods suffered from non-uniform illumination, projection of the light source on the imaging plane etc. Köhler illumination eliminates non-uniform illumination so that all parts of the light source contribute to specimen illumination. It works by ensuring that the lamp image is not projected on the sample plane with the use of a collector lens placed near the lamp. This lens focuses the image of the lamp to the condenser lens. Under this condition, illumination of the specimen is uniform.

The specimen stage is used for placing the specimens under examination. The stage can be adjusted to move along its two axes, so that a large specimen can be imaged. Depending on the features of a microscope, the stage could be manual or motor controlled.

Focus knobs allow moving the stage or objective in the vertical axis. This allows focusing of the specimen and also enables imaging of large objects.

12.4 Epi-Illumination

In the microscope setup shown in Figure 12.1, the specimen is illuminated by using a light source placed below. This is called transillumination. This method does not separate the emission and excitation light in fluorescence microscopy. An alternate method called epi-illumination is used in modern microscopes.

In this method (Figure 12.7), the light source is placed above the specimen. The dichoric mirror reflects the excitation light and illuminates the specimen. The emitted light (which is of longer wavelength) travels through the dichoric mirror and is either viewed or detected using a camera. Since there are two clearly defined paths for emission and excitation light, only the emitted light is used in the formation of the image and hence improves the quality of the image.

12.5 Fluorescence Microscope

A fluorescence microscope allows identification of various parts of the specimen not only in terms of structure but also in terms of function. It allows tagging different parts of the specimen, so that it generates light of certain wavelengths and form an image.

12.5.1 Theory

Fluorescence is generally observed when a fluorescent molecule absorbs light at a particular wavelength and emits light at a different wavelength within a short interval. The molecule is generally referred to as fluorochrome or dye, and the delay between absorption and emission is in the order of nanoseconds. This process is generally shown diagrammatically using the Jablonski diagram shown in Figure 12.5.

The figure should be read from bottom to top. The lower state is the stable ground state, generally called S_0 state. A light or photon incident on the fluorochrome causes the molecule to reach an excited state (S_1'). A molecule in the excited state is not stable and hence returns back to its stable state after losing the energy both in the form of radiation such as heat and also light of longer wavelength. This light is referred to as the emitted light.

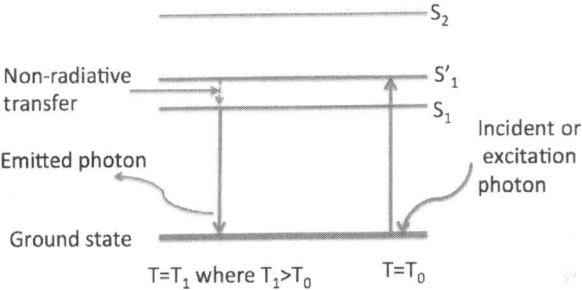

FIGURE 12.5: Jablonski diagram.

From Planck's law that was discussed in Chapter 10, X-Ray and Computed Tomography, the energy of light is inversely proportional to the wavelength. Thus, light of higher energy will have shorter wavelength and vice versa. The incident photon is a higher energy and hence shorter wavelength while the emitted light is of low energy and longer wavelength. The exact mechanisms of emission and absorption are beyond the scope of this book and readers are advised to consult books dedicated to fluorescence for details.

12.5.2 Properties of Fluorochromes

Two properties of fluorochromes, excitation wavelength and emission wavelength, were discussed in the previous section. Table 12.2 lists the excitation and emission wavelengths of commonly used fluorochromes. As can be seen in the table, the difference between excitation and emission wavelengths, or the Stokes shift, is significantly

Fluorochrome	Peak Excitation Wavelength (nm)	Peak Emission Wavelength (nm)	Stokes Shift (nm)
DAPI	358	460	102
FITC	490	520	30
Alexa Fluor 647	650	670	20
Lucifer Yellow VS	430	536	106

TABLE 12.2: List of the fluorophores of interest to fluorescence imaging.

different for different dyes. The larger the difference, the easier it is to filter the signal between emission and excitation.

A third property, quantum yield, is also important in characterizing dye. It is defined as:

$$QY = \frac{\text{Number of emitted photons}}{\text{Number of absorbed photons}} \tag{12.6}$$

Another important property that determines the amount of fluorescence generated is the absorption cross-section. The absorption cross-section can be explained with the following analogy. If a bullet is fired at a target, the ability to reach the target is easier if the target is large and if the target surface is oriented in the direction perpendicular to the direction of bullet path. Similarly, the term absorption cross-section defines the "effective" cross-section of the fluorophore and hence the ability of the excitation light to produce fluorescence.

It is measured by exciting a sample of fluorophore of certain thickness with excitation photon of a certain intensity and measuring the intensity of the emitted light. The relationship between the two intensities is given by Equation 12.7.

$$I = I_0 e^{-\sigma D \delta x} \tag{12.7}$$

where I_0 is the excitation photon intensity, I is the emitted photon intensity, σ is the absorption cross-section of the fluorophore, D is the density, and δx is the thickness of the fluophore.

12.5.3 Filters

During fluorescence imaging, it is necessary to block all light that is not emitted by the fluorochrome. This ensures the best contrast in the image and consequently better detection and image processing. In addition, the specimen does not necessarily contain only one type of fluorochrome. Thus, to separate the image created by one fluorochrome from the other, a filter that allows only light of a certain wavelength corresponding to the different fluorochromes is needed.

The filters can be classified into three categories: lowpass, bandpass and highpass. The lowpass filter allows light of shorter wavelength and blocks longer wavelengths. The highpass filter allows light of longer wavelength and blocks shorter wavelengths. The bandpass filter allows light of a certain range of wavelengths. In addition, fluorescence microscopy uses a special type of filter called a dichroic mirror (Figure 12.7). Unlike the three filters discussed earlier, in a dichroic mirror the incident light is at 45^0 to the filter. The mirror reflects light of shorter wavelength and allows longer wavelength to pass through.

Multi-channel imaging is a mode where different types of fluorochromes are used for imaging resulting in images with multiple channels. Such images are called multi-channel images. Each channel contains image corresponding to one fluorochrome. For example if we obtained an image of size 512-by-512, using two different fluorochromes, the image would be of size 512-by-512-by-2. The two in the size corresponds to the two channels. Generally most fluorescence images have 3 dimensions. Hence the volume in such cases would be 512-by-512-by-200-by-2, where 200 is the number of slices or z-dimension. The actual number may vary based on the imaging conditions.

The choice of the fluorochrome is dependent on the following parameters:

1. Excitation wavelength

2. Emission wavelength

3. Quantum yield

4. Photostability

Filters used in the microscope need to be chosen based on the fluorochrome being imaged.

12.6 Confocal Microscopes

Confocal microscopes overcome the issue of spatial resolution that affects wide-field microscopes. A better resolution in confocal microscopes is achieved by the following:

- A narrow beam of light illuminates a region of the specimen. This eliminates collection of light by the reflection or fluorescence due to a nearby region in the specimen.

- The emitted or reflected light arising from the specimen passes through a narrow aperture. A light emanating from the direction of the beam will pass through the aperture. Any light emanating from nearby objects or any scattered light from various objects in the specimen will not pass through the aperture. This process eliminates all out-of-focus light and collects only light in the focal plane.

The above process describes image formation at a single pixel. Since an image of the complete specimen needs to be formed, the narrow beam of light needs to be scanned all across the specimen and the emitted or reflected light needs to be collected to form a complete image. The scanning process is similar to the raster scanning process used in television. It can be operated using two methods. In the first method devised by Marvin Minsky, the specimen is translated so that all points can be scanned. This method is slow and also changes the shape of the specimen suspended in liquids and is no longer used. The second approach is to keep the specimen stationary while the light beam is scanned across the specimen. This was made possible by advances in optics and computer hardware and software, and is used in all modern microscopes.

12.7 Nipkow Disk Microscopes

Paul Nipkow created and patented a method for converting an image into an electrical signal in 1884. The method consisted of scanning an image by using a spinning wheel containing holes placed in a spiral pattern, as shown in Figure 12.6. The portion of the wheel that does not contain the hole is darkened so that light does not pass through it. By rotating the disk at constant speed, a light passing through the hole scanned all points in the specimen. This approach was later adopted to microscopy. Figure 12.6 shows only one spiral with a smaller number of holes while a commercially available disc will have large number of holes, to allow fast image acquisition.

A setup containing the disk along with the laser source, objective lens, detector and the specimen is shown in Figure 12.7, and Figure 12.8 is a photograph of a Nipkow disk microscope. In this figure, the illuminating light floods a significant portion of the holes. The portion that does not contain any holes reflects the light. The light that passes through the holes reaches the specimen through the objective lens. The reflected light, or the light emitted by fluorescence, passes through the objective and is reflected by the dichroic mirror. The detector forms an image using the reflected light.

Unlike a regular confocal microscope, the Nipkow disk microscope

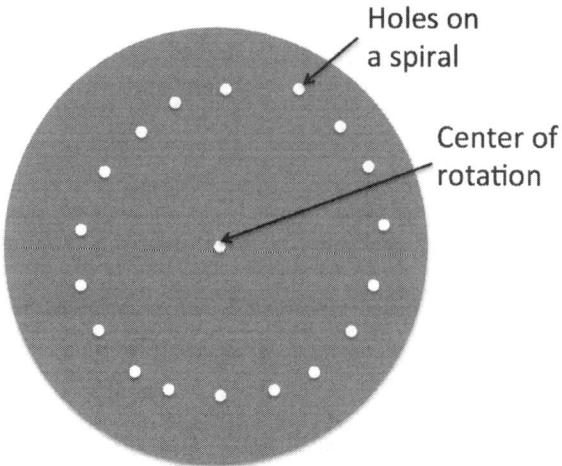

FIGURE 12.6: Nipkow disk design.

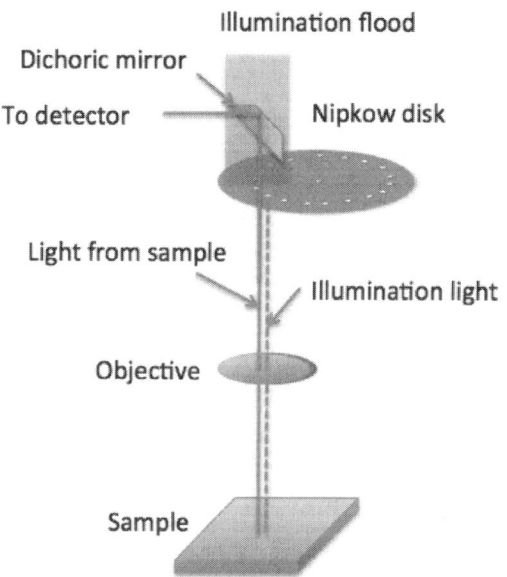

FIGURE 12.7: Nipkow disk setup.

is faster as neither the specimen nor the light beam needs to be raster scanned. This enables rapid imaging of live cells.

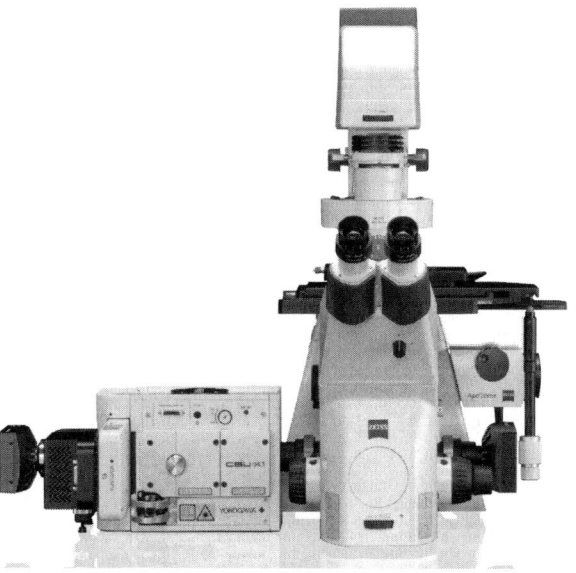

FIGURE 12.8: Photograph of Nipkow disk microscope. Original image reprinted with permission from Carl Zeiss Microscopy, LLC.

12.8 Confocal or Wide-Field?

Confocal and wide-field microscopes each have their own advantages and disadvantages. These factors need to be considered when making a decision on what microscope to use for a given cost or type of specimen.

- Resolution: There are two different resolutions: xy and z direction. Confocal microscopes produce better resolution images in both directions. Due to advances in computing and better software, wide-field images can be deconvolved to a good resolution along xy but not necessarily along the z direction.

- Photo bleaching: Images from a confocal microscope may be photo-bleached, as the specimen is imaged over a longer time period compared to a wide-field microscope.

- Noise: Wide-field microscopes generally produce images with less noise.

- Acquisition rate: Since confocal images scan individual points, it is generally slower compared to wide-field microscope.

- Cost: As a wide-field microscope has fewer parts, it is less expensive than confocal.

- Computer processing: Confocal images need not be processed using deconvolution. Depending on the setup, deconvolution of a wide-field image can produce images of comparable quality to confocal images.

- Specimen composition: A wide-field microscope with deconvolution works well for a specimen with a small structure.

12.9 Summary

- The physical properties that govern optical microscope imaging are magnification, diffraction limits, and numerical aperture. The diffraction limit and numerical aperture determine the resolution of the image.

- The specimen is immersed in a medium in order to match the refractive index and also to increase the resolution.

- Wide-field and confocal are the two most commonly used microscopes. In the former, a flood of light is used to illuminate the specimen while in the latter, a pencil beam is used to scan the specimen and the collected light passes through a confocal aperture.

- The fluorescence microscope allows imaging of the shape and function of the specimen. Fluorescence microscope images are obtained after the specimen has been treated with a fluorophore.

- The specific range of wavelength emitted by the fluorophore is measured by passing the light through a filter.

- To speed up confocal image acquisition, a Nipkow disk is used. The disk consists of a series of holes placed on a spiral. The disk is rotated and the position of the holes is designed to ensure that complete 2D scanning of the specimen is achievable.

12.10 Exercises

1. If the objective has a magnification of 20X and the eyepiece has a magnification of 10X, what is the total magnification?

2. A turret has three objectives: 20X, 40X and 50X. The eyepiece has magnification of 10X. What is the highest magnification?

3. In the same turret setup, if a cell occupies 10% of the field of view for an objective magnification of 20X, what would be the field of view percentage for 40X?

4. Discuss a few methods to increase spatial resolution in an optical microscope. What are the limits for each parameter?

Chapter 13

Electron Microscopes

13.1 Introduction

The resolution of a light microscope discussed previously is directly proportional to the wavelength. To improve the resolution, light with a shorter wavelength should be used. Scientists began experimenting with ultraviolet light, which has a shorter wavelength than visible light. Due to the difficulty in generation and maintaining coherence, it was not commercially successful.

Meanwhile, the French physicist Louis de Broglie proved that similar to visible light, a traveling electron has both wave and particle duality. He was awarded a Nobel Prize in 1929 for his work.

An electron wave with higher energy will have lower wavelength and vice versa. Thus, improving the resolution would involve increasing the energy. The wavelength of electrons is considerably shorter than that of visible light and hence very high-resolution images can be obtained. Visible light has a wavelength of 400 - 700 nm. Electrons, on the other hand, have a wavelength of 0.0122 nm for an accelerating voltage of 10 kV.

Ernst Ruska and Max Knoll created the first electron microscope (EM) with the ability to magnify objects 400 times. Upon further work, Ruska improved its resolution beyond the resolution of optical microscopes and hence made the EM an indispensable tool for microscopists. The EM used today does not measure a single characteristic property but rather measures multiple characteristics of the material. The one common thing among all of them is the electron beam.

In Section 13.2, we discuss some of the physical principles that need to be understood regarding EM. We begin with a discussion on the properties of the electron beam and its ability to produce images at high resolution. We introduce the interaction of electrons with the matter and various particles and waves that are generated as a consequence. The fast moving electron beam from the electron gun passes through the material to be imaged. During its transit through the material, the electron interacts with the atoms in that material. We integrate the two basic principles and discuss the construction of an EM. We also discuss specimen preparation and general precautions to be carried out when preparing the material. The electron beam has to travel through a vacuum before it reaches the material. Hence the material has to be placed in the vacuum chamber along with other components. This limits the type of material that can be imaged. Some materials have to be prepared for EM image acquisition. Interested readers can refer to [6],[21],[32],[27],[30],[31],[47],[49],[50],[85],[101],[105].

13.2 Physical Principles

The EM was made possible by many fundamental and practical discoveries made over time. In this section, we discuss these discoveries and place them in the context of creating an electron microscope.

EM process involves bombarding a high-speed electron beam on the specimen and recording the beam emanating from or transmitted through the specimen. These high-speed electrons have to be focused to a point in the specimen and also navigate to all other points. In 1927, Hans Busch proved that an electron beam can be focused on an inhomogeneous magnetic field just as light can be focused using a lens.

Four years later, Ernst Ruska and Max Knoll confirmed this by constructing such a magnetic lens. This lens is still a part of today's EM design.

The second basic principle is the dual nature of the electron beam proven by Louis de Broglie. The electron beam behaves as wave and particle just like visible light. Thus the electron beam has both wavelength and mass.

13.2.1 Electron Beam

Louis de Broglie proved that electrons traveling at high speed have both particle and wave natures. The wavelength of the beam is given by Equation 13.4. Thus the faster the electrons travel the lower is the wavelength of the beam. As we will discuss later, the lower wavelength results in production of high-resolution images.

$$\lambda = \frac{h}{m\nu} \tag{13.1}$$

where h is the Planck's constant and is equal to $6.626\ 10^{-34}$ Js, m is the electron mass and is equal to $m = 9.109\ 10^{-31}$ kg and ν is the frequency.

We also know that the beam stores the energy in the form of kinetic energy given by the following equation.

$$E = \frac{mv^2}{2} = eV \tag{13.2}$$

where $e = 1.602\ 10^{-19}$ coulombs is the charge of the electron and V is the acceleration voltage. In other words,

$$v = \sqrt{\frac{2eV}{m}} \tag{13.3}$$

Plug in this equation into Equation 13.1 to obtain

$$\lambda = \frac{h}{\sqrt{2meV}} \tag{13.4}$$

Since all the variables on the right-hand side of the equation are constant except for the accelerating voltage V, we can simplify the equation to

$$\lambda = \frac{1.22}{\sqrt{V}} \qquad \text{nano-meter} \qquad (13.5)$$

where V is voltage measured in volts. Thus for an accelerating voltage of 10 kV, the wavelength of the electron beam is 0.0122 nm.

Since the speed of the beam and the acceleration voltage are generally very high for electron microscopy, the wavelength computation is dependent on the relativistic effect.

13.2.2 Interaction of Electron with Matter

In Chapter 10, X-Ray and Computed Tomography we discussed the interaction of x-rays with materials. We discussed the Bremsstrahlung spectrum (braking spectrum) and the characteristic spectrum. The former is created as the incident x-ray is slowed by its passage through the material. The latter is formed when the x-rays knock out electrons from their orbit.

The electron beam has both particle and wave natures similar to x-rays. Hence the electron beam exhibits a spectrum similar to x-rays. Since the energy of the electron is higher than that of the x-ray, it also produces few other emissions. The various emissions are transmitted electrons, back-scattered electrons (BSE), secondary electrons (SE), elastically scattered electrons, inelastically scattered electrons, Auger electrons (AE), characteristic x-rays, Bremsstrahlung x-rays, visible light (cathodoluminesence), diffracted electrons (DE) and heat. This phenomena is shown in Figure 13.1. The various emissions occur at different depths of the material. The region that generates these emissions is referred to as electron interaction volume. SE are generated at the top of the region while the Bremsstrahlung x-rays are generated at the bottom.

In a typical EM, not all of these are measured. For example in the transmission EM or TEM, the transmitted electron, elastically scattered electron and inelastically scattered electrons are measured, and in the scanning EM (SEM), BSE or SE are measured.

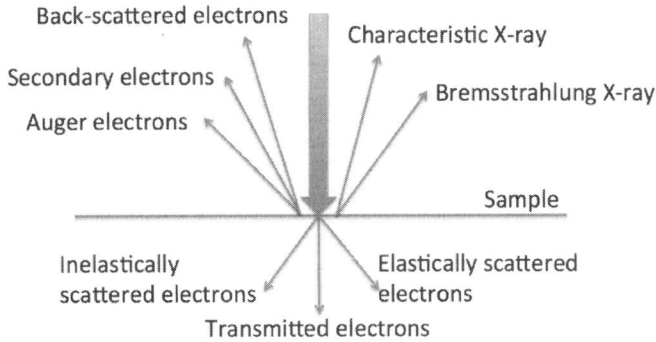

FIGURE 13.1: Intensity distributions.

Since we discussed Bremsstrahlung and characteristic x-rays earlier, we will focus on the other important emissions, the BSE, SE and TE, in this chapter.

13.2.3 Interaction of Electrons in TEM

TEM measures three different electrons during its imaging process. They are transmitted electrons, elastically scattered (or diffracted electrons), and inelastically scattered electrons.

In Chapter 10, we discussed image formation by exposing a photographic plate or a digital detector to an x-ray beam after it passes through material. The image is formed using the varying intensity of the x-ray in proportion to the thickness and attenuation coefficient of the material at various points. In TEM, the incident beam of electrons replaces the x-ray. This beam is transmitted through the specimen without any significant change in intensity, unlike x-ray. This is due to the fact that the electron beam has very high energy and that the specimen is extremely thin (on the order of 100 microns). The region in the specimen that is opaque will transmit fewer electrons and appear darker.

A part of the beam is scattered elastically (i.e., with no loss of energy) by the atoms in the specimen. These electrons follow the Bragg's law of diffraction. The resultant image is a diffraction pattern.

The inelastically scattered electrons (i.e., with loss of energy) contribute to the background. The specimen used in TEM is generally very thin. Increasing the thickness of the specimen results in more inelastic scattering and hence more background.

13.2.4 Interaction of Electrons in SEM

The TEM specimen is generally thin and hence there are fewer modes of interaction. The SEM, on the other hand, uses a thick or bulk specimen and hence has more modes of interaction in addition to the modes discussed in Section 13.2.3.

In SEM, the various modes of interaction are:

1. Characteristic x-rays

2. Bremsstrahlung x-rays

3. Back-scattered electrons (BSE)

4. Secondary electrons (SE)

5. Auger electrons

6. Visible light

7. Heat

The generation of characteristic x-rays and Bremsstrahlung x-rays was discussed in Chapter 10. The former is produced by the knock-out of an electron from its orbit by the fast moving electron while the latter is produced by deceleration of the electron during its transit through material.

The mechanism of generation of Auger electrons is similar to characteristic x-rays. When a fast-moving electron ejects an electron in orbit, it leaves a vacancy in the inner shell. An electron from a higher shell fills this vacancy. The excess energy is released as an x-ray in the case of the characteristic x-ray, while an electron is ejected during Auger electron formation. Since the Auger electron has low energy, it is generally formed only on the surface of the specimen.

SE are low voltage electrons. They are generally less than 50 eV in energy. They are generally emitted at the top of the specimen, as their energy is too small to be emitted inside the material and still escape to be detected. Since SE are emitted from the top of the surface, they are used for imaging the topography of the sample.

BSE are obtained by the scattering of the primary electron by the specimen. This scattering occurs at depths higher than the regions where SE are generated. Materials with high atomic numbers produce a significantly larger number of BSE and hence appear brighter in the BSE detector image. Since BSE are emitted from the inside of the specimen, they are used for imaging the chemical composition of the specimen and also for topographic imaging.

13.3 Construction of EM

13.3.1 Electron Gun

The electron gun generates an accelerated beam of electron. There are two different types of electron gun: thermionic gun and field emission gun. In the former, electrons are emitted by heating a filament while in the latter, electrons are emitted by the application of an extraction potential.

A schematic of the thermionic gun is shown in Figure 13.2. The filament is heated by passing current, which generates electrons by a process of thermionic emission. It is defined as emission of electrons by absorption of thermal energy. The number of electrons produced is proportional to the current through the filament. The Wehnlet cap is maintained at a small negative potential, so that the negatively charged electrons are accelerated in the direction shown through the small opening. The anode is maintained at a positive potential, so that the electrons travel down the column towards the specimen. The acceleration is achieved by the voltage between the cap and the anode.

The filament can be made of tungsten or lanthanum hexaboride (LaB_6) crystals. Tungsten filaments can work at high temperatures but do not produce circular spots. The LaB_6 crystals on the other hand can produce circular spots and hence better spatial resolution.

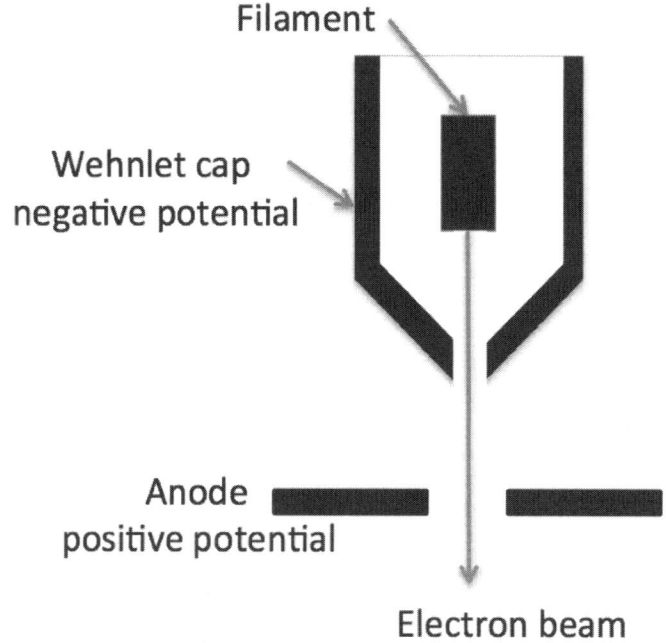

FIGURE 13.2: Thermionic gun.

A schematic of the field emission gun (FEG) is shown in Figure 13.3. The filament used is a sharp tungsten metal tip. The tip is sharpened to have a dimension on the order of 100 nm. In a cold FEG, the electron from the outer shell is extracted by using the extraction voltage (V_E). The extracted electrons are accelerated using the accelerating voltage (V_A). In the thermionic FEG, the filament is heated to generate electrons. The extracted electrons are accelerated to high energy.

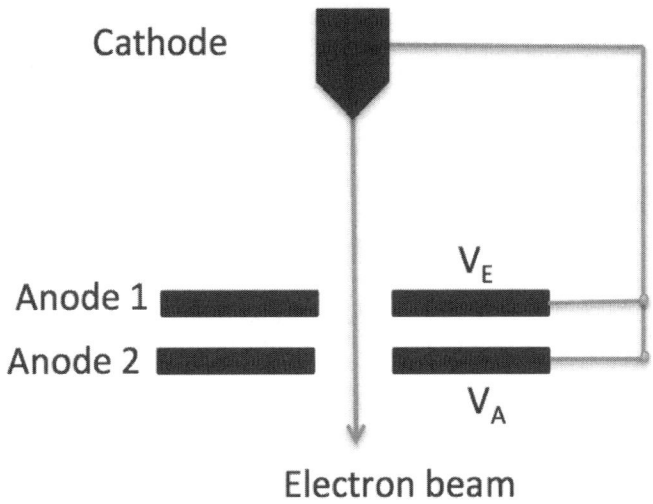

FIGURE 13.3: Field emission gun.

13.3.2 Electromagnetic Lens

In Chapter 12, Light Microscopes, we discussed the purpose of the various lenses, objective and eyepiece. The lens is chosen such that the light from the object can be focused to form an image. Since electrons behave like waves, they can be focused by using lenses.

From our discussion of Image Intensifier (II) in Chapter 10, the electrons are affected by the magnetic field. In the case of II, this phenomenon presents a problem and results in distortion. However, a controlled magnetic field can be used to navigate electrons and hence create a lens. It has been proven that an electron traveling through vacuum in a magnetic field will follow a helical path.

The electrons enter the magnetic field at point O1 (Figure 13.4). Point O2 is the point where all electrons generated by the electron gun are focused by the magnetic field. The distance O1-O2 is the focal

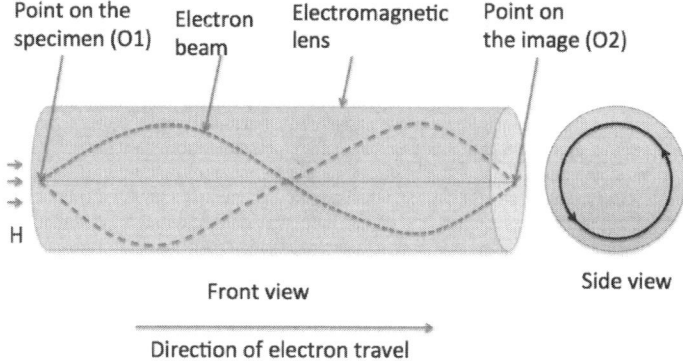

FIGURE 13.4: Electromagnetic lens.

length of the lens. The mathematical relationship that defines focal length is given by

$$f = K\frac{V}{i^2} \qquad (13.6)$$

where K is a constant based on the design of the coil at the geometry, V is the accelerating voltage and i is the current through the coil. As can be seen, either increasing the voltage or reducing the current in the coil can increase focal length. In an optical microscope, the focal length for a given lens is fixed while it can be changed in the case of an electromagnetic lens. Hence, in an optical microscope, the only method for changing the focal length is by either changing the lens (using objective turret) or by changing spacing between the lenses. On the other hand, in an electromagnetic lens, the magnification can be changed by altering the voltage and current. The electromagnetic lens suffers from aberrations similar to optical lenses. Some of these are astigmatism, chromatic aberration, and spherical aberration. They can be overcome by designing and manufacturing under high tolerance.

13.3.3 Detectors

Secondary electron detectors: SE are measured using the Everhart-Thornley detector (Figure 13.5). It consists of a Faraday's cage, a scintillator, a light guide and a photo-multiplier tube. SE have very low energy (less than 50 eV). To attract these low energy electrons, a positive voltage on the order of 100 V is applied to the Faraday's cage in order to attract SE. The scintillator is maintained at a very high positive voltage to attract the SE. The SE are converted to light photons by the scintillator. The light generated is too weak to form an image. Hence the light is guided through a light guide onto a photo-multiplier tube, which amplifies the light signal to form an image.

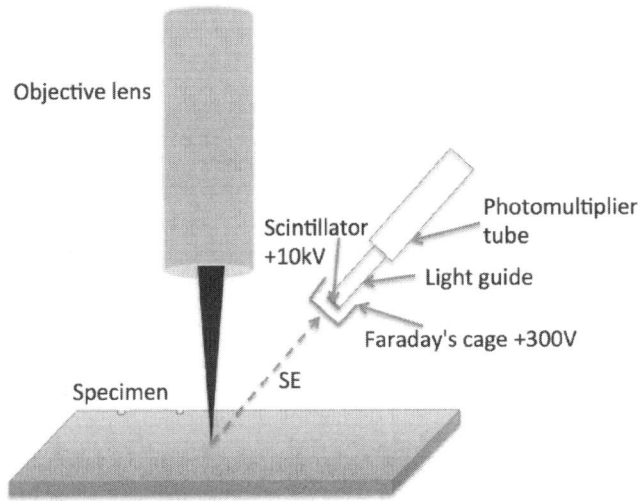

FIGURE 13.5: Everhart-Thornley secondary electron detector.

Back-scattered electron detectors: BSE have very high energy and hence readily travel to a detector. BSE also travel in all directions and hence a directional detector such as the Everhart-Thornley detector can only collect a few electrons and will not be enough to form a complete image. BSE detectors are generally doughnut shaped (Figure 13.6) and placed around the electron column just below the objective lens. The detecting element is either a semiconductor or a scintillator, that converts the incident electron into light photons that are recorded using a camera.

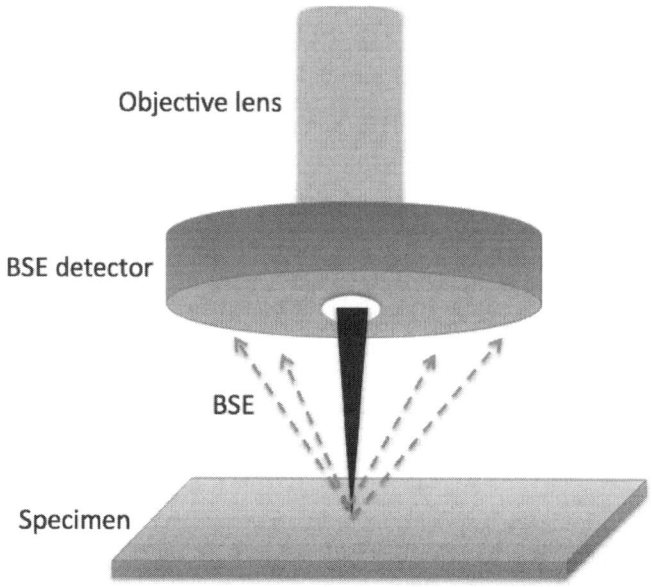

FIGURE 13.6: Back-scattered electron detector.

13.4 Specimen Preparations

The specimen needs to be electrically conductive. Hence biological specimens are coated with a thin layer of electrically conductive material such as gold, platinum, tungsten etc. In some cases, a biological sample cannot be coated without affecting the integrity of the specimen. In such cases, SEM can be operated at low voltage. Materials made of metal do not have to be coated, as they are electrically conductive.

Since the electron beam can only travel in vacuum, the specimen needs to be prepared for placement in a vacuum chamber. Materials with water need to be dehydrated. The dehydration process causes the specimen to shrink and change in shape. Hence the specimen has to be chemically fixed in which water is replaced by organic compounds. The specimen is then coated with electrically conductive material before imaging.

An alternate method is to freeze the sample using cryofixation. In this method, the specimen is cooled rapidly by plunging into liquid nitrogen (boiling point $= -195.8^oC$). The rapid cooling of the specimen preserves its internal structure so that it can be imaged accurately. The rapid cooling ensures that ice crystals, which can damage the specimen, do not form. In the case of TEM, since the specimen has to be thin, the cryofixated specimen is cut into thin slices or microtomy.

13.5 Construction of TEM

In the previous sections, we have discussed the various components of TEM and SEM. In the next two sections, we will integrate the various parts to construct TEM and SEM. Figure 13.7 illustrates the barebones optical microscope, TEM and SEM. Although this discussion is for illustration purposes, the complete equipment consists of multiple controls to ensure good image quality.

In each case, a source of light or electron is at the top. The light in the case of the optical microscope travels through a condenser lens, a specimen and then the objective or eyepiece to be either viewed by an eye or imaged using a detector.

In the case of TEM, the source is the electron gun. The accelerated electrons are focused using a condenser lens, transmitted through the specimen and finally focused to form an image using objective and eyepiece magnetic lenses. Since the electron beam can only travel in vacuum, the entire setup is placed in a vacuum chamber.

An example of an image of Sindbis virus obtained using TEM is shown in Figure 13.8 [108].

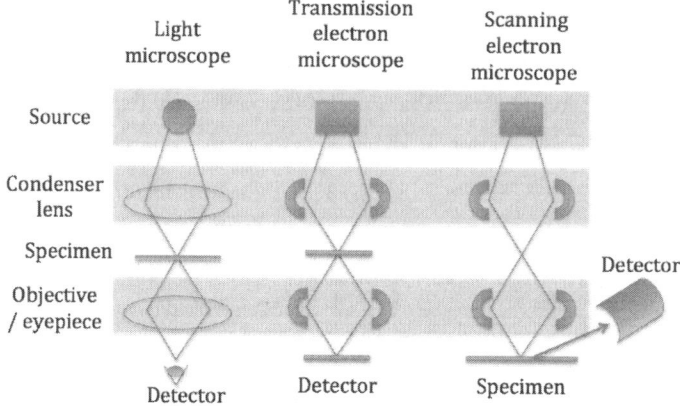

FIGURE 13.7: Comparison of optical microscope, TEM and SEM.

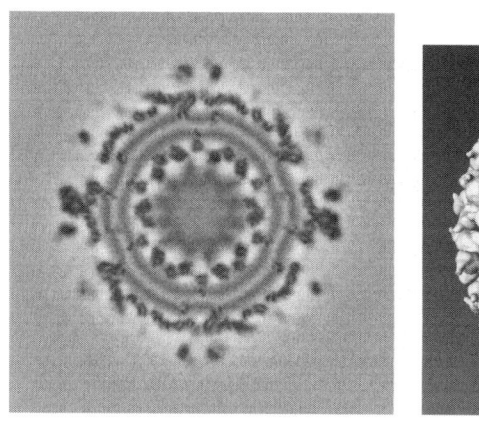
(a) A slice of 3D image obtained using a TEM.

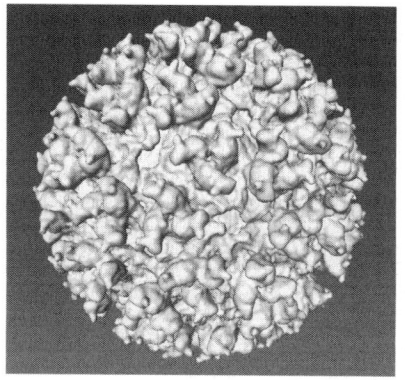
(b) All slices rendered to an iso-surface.

FIGURE 13.8: TEM slice and its iso-surface rendering. Original image reprinted with permission from Dr. Wei Zhang, University of Minnesota.

13.6 Construction of SEM

Figure 13.7 illustrates the schematics of light, TEM and SEM. Figure 13.9 is an example of an SEM machine. In the case of SEM, the source is the electron gun as in TEM. The accelerated electron is then focused using a condenser lens to form a small spot on the specimen. The electron beam interacts with the specimen and emits BSE, SE, Auger electrons etc. These are measured using the detector discussed previously to form an image. Since the electron beam can only travel in vacuum, the entire setup is placed in a vacuum chamber.

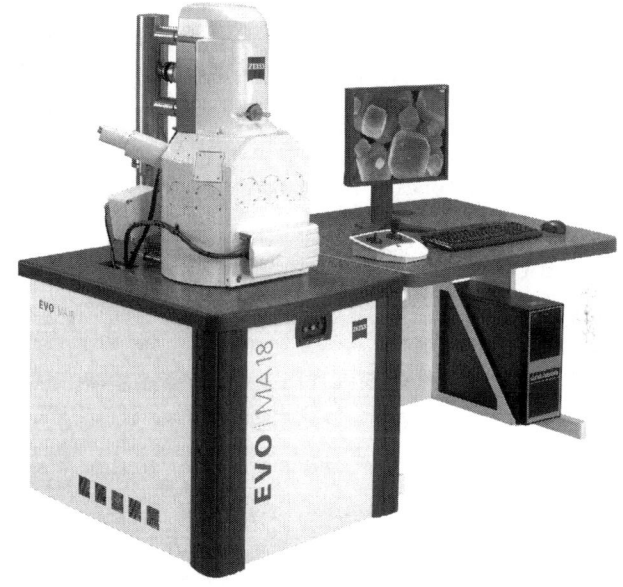

FIGURE 13.9: An SEM machine. Original image reprinted with permission from Carl Zeiss Microscopy, LLC.

An example of an image obtained using SEM is shown in Figure 13.10.

13.7 Summary

- EM involves bombarding high-speed electron beams on a specimen and recording its response.

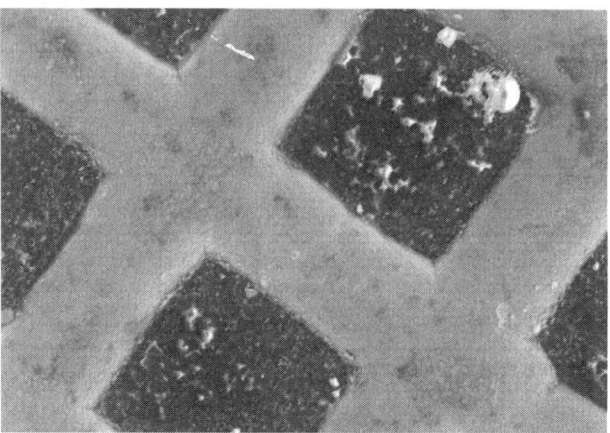

FIGURE 13.10: BSE image obtained using an SEM.

- Imaging an electron is possible, as it exhibits both particle and wave natures.

- The wavelength of the electron is inversely proportional to the square root of the accelerating voltage. Increasing the accelerating voltage results in lower wavelength or higher resolution. The typical accelerating voltage is 30kV.

- A high-speed electron beam bombards a specimen and generates characteristic x-rays, Bremsstrahlung x-ray, back-scattered electrons (BSE), secondary electrons (SE), Auger electrons, visible light, and heat. BSE and SE are the most commonly measured in SEM.

- The EM focuses the beam using an electromagnetic lens.

- The BSE is measured using a doughnut shaped detector wrapped around the axis of the electron beam.

- The SE is measured using an Everhart-Thornley detector.

- Unlike a light microscope, in the case of electron microscopes, the specimen needs to be carefully prepared as the imaging is conducted in vacuum.

- The parameters that determine the quality of image are voltage, working distance, and spot size.

13.8 Exercises

1. The accelerating voltage of an SEM is 10kV. Calculate the wavelength of the generated electron.

2. Compare and contrast the working principles of TEM and SEM.

3. List the order of generation of various spectrums in the electron interaction volume beginning with the surface of the specimen.

Appendix A

Installing Python Distributions

In Chapter 1, Introduction to Python, we briefly discussed various Python distributions. Operating systems such as MacOSX and Linux come prebuilt with a Python interpreter. This Python interpreter is not ready for scientific computation as it does not have the common scientific modules such as numpy, scipy etc. Installing these modules requires knowledge of compiling complex mathematical libraries such as MKL, Boost etc. Various distributions have been created to ease the task of installing a Python distribution for scientific computation. Some of these distributions are free while others are free only for academic community users. We will discuss in detail the two most popular distributions and methods for installing them on your favorite OS. The two distributions are Enthought Python distribution (EPD) and PythonXY. The former is available for MacOSX, Windows and Linux while the latter is available only for Windows.

PythonXY also installs "Spyder", an interface for Python programming created to be similar to the MATLAB®interface. EPD does not install Spyder by default but it can be configured.

A.1 Windows

A.1.1 PythonXY

The installer can be downloaded from http://www.pythonxy.com. Once downloaded, double-click to begin installing. The installation process consists of multiple pages which can be navigated by using the "Next" button. Only the most important pages will be discussed. In Figure A.1 install types are specified. In this case, the install type was chosen to be full i.e., all the plug-ins and modules will be installed. The other pages can be navigated using the "Next" button.

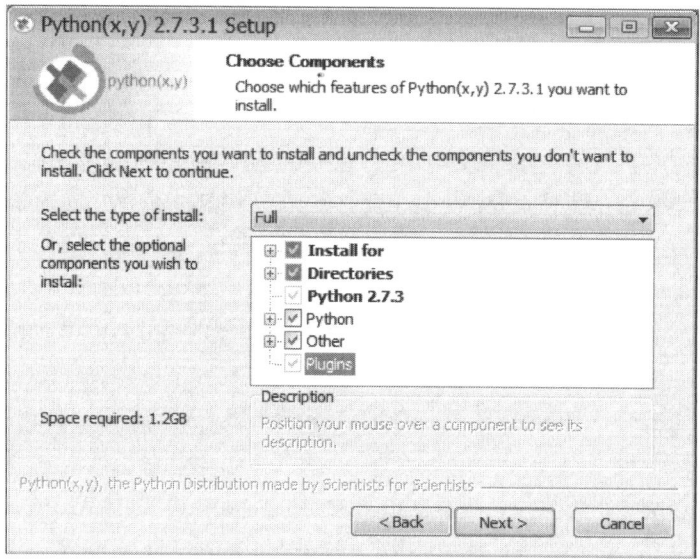

FIGURE A.1: Specifying the type of install.

The interpreter can be accessed through Spyder from the windows menu shown in Figure A.2. The Spyder interface is shown in Figure A.3. The interface is designed to look similar to MATLAB interface. The left column is the editor for creating a Python program. The right column consist of two sections. The top section contains the object inspector, variable explorer and the file explorer. The object inspector provides interactive documentation of any Python function. The variable explorer lists all the variables that are currently used in the Python interpreter. The file explorer allows easy navigation of all the files in the folder in the system. The bottom section consists of the console and history log. The console is a convenient place for testing Python functions before they are incorporated into a program. History log consists of all the commands that were typed in the console.

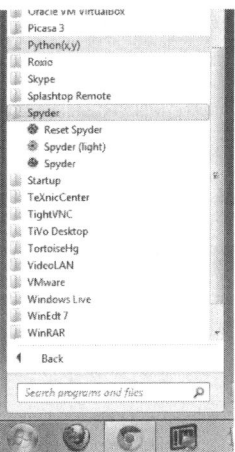

FIGURE A.2: The Windows menu item to start PythonXY under Spyder.

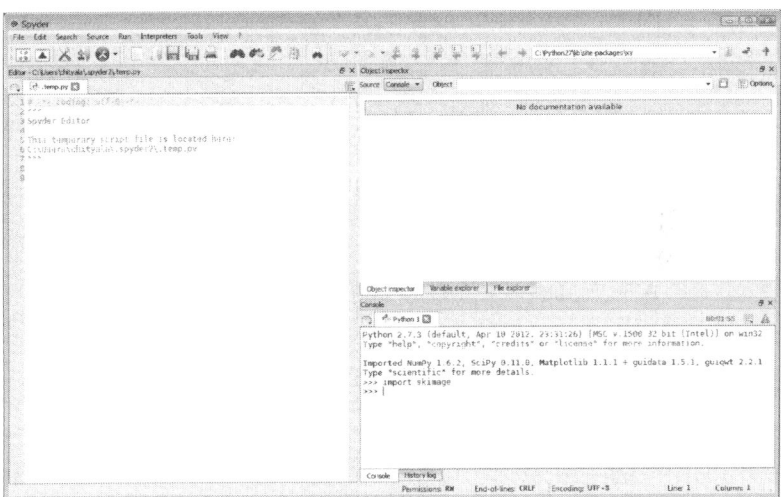

FIGURE A.3: The Spyder interface.

A.1.2 Enthought Python Distribution

The installer can be downloaded from http://enthought.com/. Once downloaded, double-click to begin installation. The installation process consists of multiple pages that can be navigated by using the "Next" button. In Figure A.4, the installer specifies the location of an interpreter that's already installed. In the absence of any interpreter installed previously, this page will not be displayed.

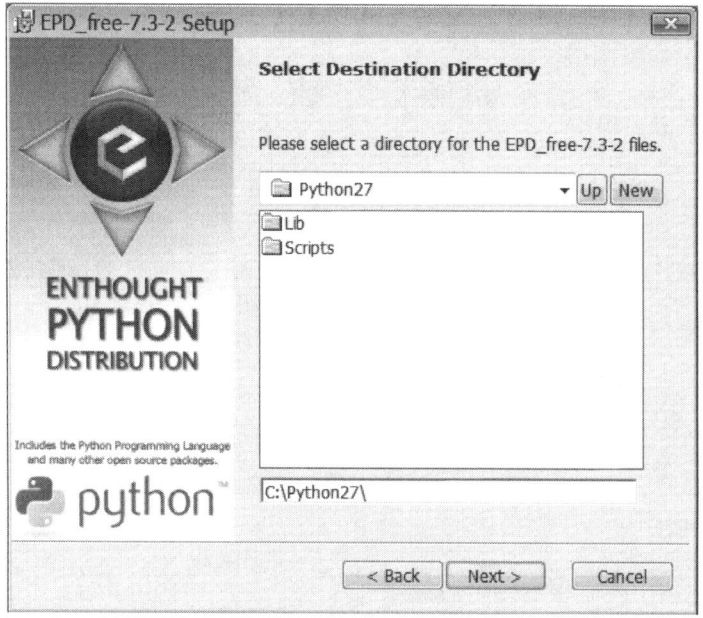

FIGURE A.4: Specifying a Python distribution for installation.

A.1.3 Updating or Installing New Modules

PythonXY and EPD are prebuilt with scientific modules. Updating to a new version of these modules is critical as a new version may contain new features, new algorithms, and bug fixes. This can be completed by downloading the appropriate installer (.exe) file from the appropriate website.

The same process can be adapted to download a module that is not

Installing Python Distributions 317

part of the Python distribution. An example of such an installation is shown in Figures A.5 and A.6. The Python module skimage is not available as part of EPD free. The appropriate module can be downloaded and installed. In Figure A.6 the installer indicates the version and the location of the interpreter to which the module will be installed.

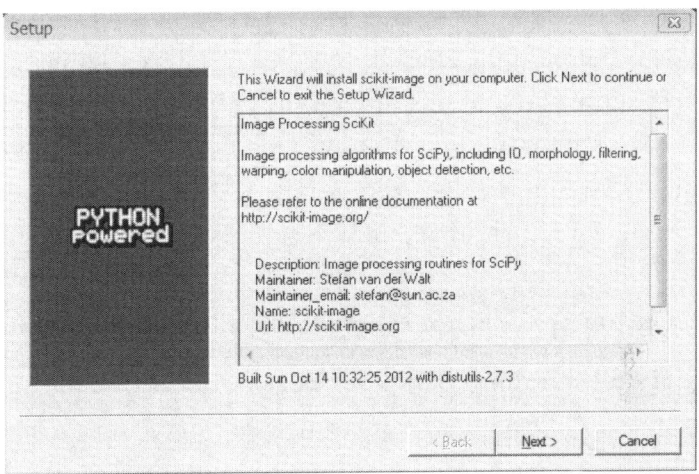

FIGURE A.5: Installation of skimage module.

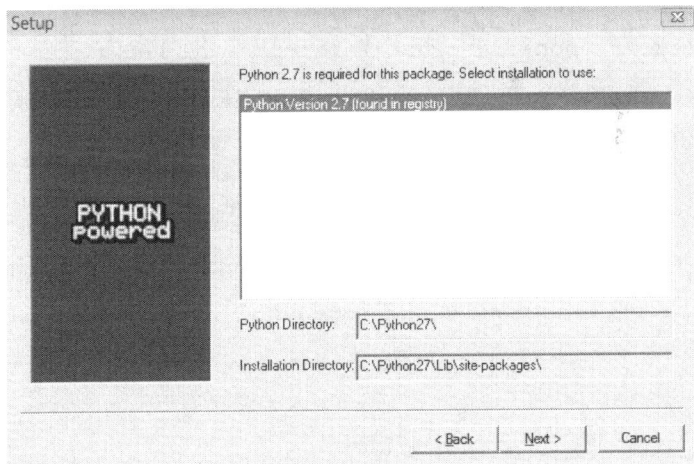

FIGURE A.6: Specifying the interpreter version and location.

A.2 Mac or Linux

Python is pre-installed in Linux and MacOSX. However, the version of Python may not contain scientific modules that are of interest to readers of this book. In this section, we will discuss installing Enthought Python distribution on a Mac. In the case of Linux, the installer is a .sh file instead of the .dmg file used in MacOSX. We will also discuss instructions for updating any Python module, using scikits-image as an example. The instructions for installing or updating any Python module is same for both Linux and Mac.

A.2.1 Enthought Python Distribution

The installer (.dmg file) can be downloaded from http://enthought.com/. Once downloaded, double-click to begin installing. The installation process consists of multiple pages that can be navigated by using the "Next" button. The installation process begins with the screen shot shown in Figure A.7. This figure indicates that version 7.3.2 of EPD free will be installed in the location /Library/Frameworks/Python.framework/Versions/7.3. This location is immutable and hence the files cannot be moved without breaking the installation.

Once the installation of EPD is complete, the Python interpreter can be invoked by typing "python" at the Mac command prompt (Figure A.8). Notice that when scikits-image module is loaded using "import skimage" command, it fails as skimage is not part of EPD free distribution.

A.2.2 Installing New Modules

New modules can be installed by using "easy_install" that comes with EPD free distribution. EasyInstall, [48] is a package manager for Python. It allows a standard method for packaging and distributing

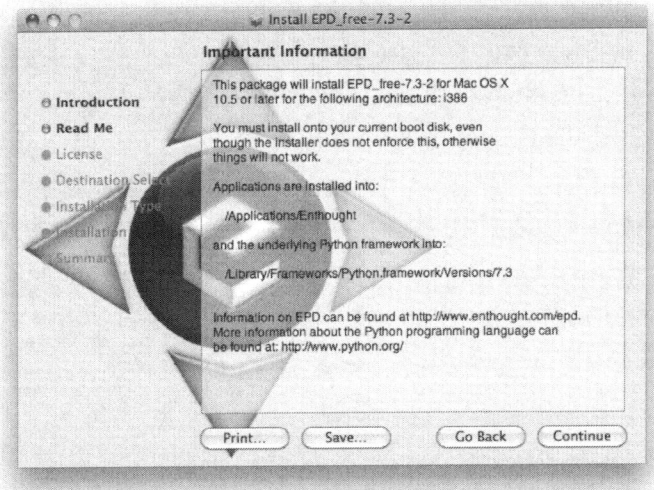

FIGURE A.7: Installing Enthought Python distribution on Mac.

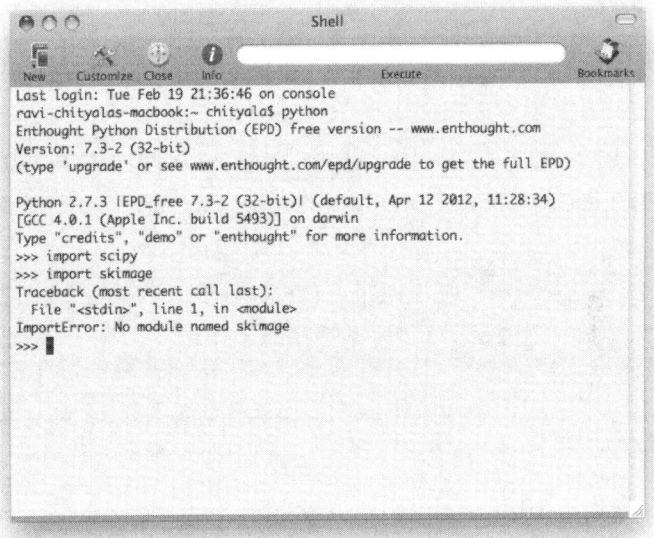

FIGURE A.8: Loading Enthought Python distribution on Mac and skimage module.

Python programs, libraries and modules. It searches the web for the packages that are requested. It begins the search with the Python Package Index (also know as pypi) and looks up metadata of locations from which the package can be downloaded. In this case, the installation of the "skimage" module requires the "cython" module, [17]. This module can be installed using "easy_install cython". The output during this process is shown in Figure A.9. As can be seen in the screenshot, the index pypi.python.org is used to obtain the location of the installer. In this case, it is `http://www.cython.org/release/Cython-0.18.zip`. EasyInstall downloads, compiles and installs the package.

```
yalas-macbook:~ chityala$ easy_install cython
Searching for cython
Reading http://pypi.python.org/simple/cython/
Reading http://www.cython.org
Reading http://cython.org
Best match: Cython 0.18
Downloading http://www.cython.org/release/Cython-0.18.zip
Processing Cython-0.18.zip
Writing /var/folders/h0/h0-Syz3bEq8hHxNsF4Ar3U+++TI/-Tmp-/easy_install-0q3aQd/Cython-0.18
/setup.cfg
Running Cython-0.18/setup.py -q bdist_egg --dist-dir /var/folders/h0/h0-Syz3bEq8hHxNsF4Ar
3U+++TI/-Tmp-/easy_install-0q3aQd/Cython-0.18/egg-dist-tmp-wPzWrg
Compiling module Cython.Plex.Scanners ...
Compiling module Cython.Plex.Actions ...
```

FIGURE A.9: Installing cython module using easy_install. This module is required to use skimage module.

After the installation of cython, skimage can be installed using "easy_install scikits-image" as shown in Figure A.10.

```
ravi-chityalas-macbook:~ chityala$ easy_install scikits-image
Searching for scikits-image
Reading http://pypi.python.org/simple/scikits-image/
Reading http://scikits-image.org
Reading http://github.com/scikits-image/scikits-image
Best match: scikits-image 0.7.1
Downloading http://pypi.python.org/packages/source/s/scikits-image/scikits-image-0.7.1.ta
r.gz#md5=d1108974cdf3eeeda0b48aafbada9c1e
Processing scikits-image-0.7.1.tar.gz
Writing /var/folders/h0/h0-Syz3bEq8hHxNsF4Ar3U+++TI/-Tmp-/easy_install-tCeb61/scikits-ima
ge-0.7.1/setup.cfg
```

FIGURE A.10: Installing skimage module using easy_install.

Finally, we can verify the installation of skimage by invoking "import skimage" in the Python interpreter (Figure A.11).

pydicom module

One of the modules that is not included in the two distributions but will be needed for reading medical images is the pydicom module.

```
Enthought Python Distribution (EPD) free version -- www.enthought.com
Version: 7.3-2 (32-bit)
(type 'upgrade' or see www.enthought.com/epd/upgrade to get the full EPD)

Python 2.7.3 |EPD_free 7.3-2 (32-bit)| (default, Apr 12 2012, 11:28:34)
[GCC 4.0.1 (Apple Inc. build 5493)] on darwin
Type "credits", "demo" or "enthought" for more information.
>>> import skimage
>>>
```

FIGURE A.11: Loading skimage module.

This module allows reading and writing dicom files. You can find more details and the link for downloading and installing pydicom at http://code.google.com/p/pydicom/.

The pydicom module can be installed on Linux and on Mac using the command "easy_install pydicom" If you do not have the required permission for pydicom on the Python installation, use virtualenv [3].

Installation in Windows can be performed by downloading the executable file. During the installation process make sure that the correct "Python directory" is provided as shown in Figure A.12.

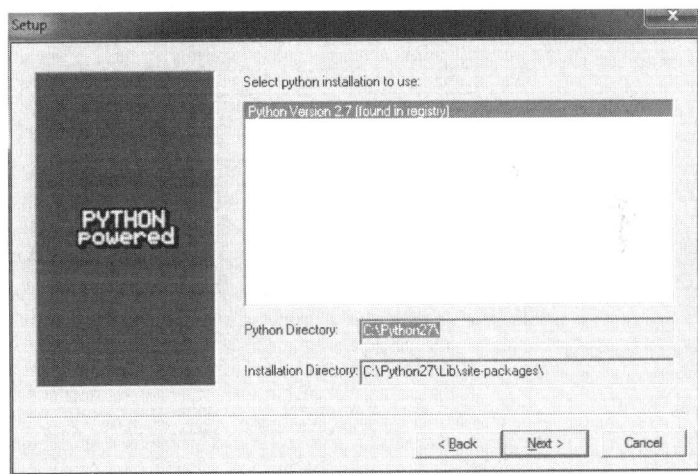

FIGURE A.12: Steps for installing pydicom on Windows.

Appendix B

Parallel Programming Using MPI4Py

B.1 Introduction to MPI

Message Passing Interface (MPI) is a system designed for programming parallel computers. It defines a library of routines that can be programmed using Fortran or C and is supported by most hardware vendors. There are popular MPI versions both free and commercially available for use. MPI version 1 was released in 1994. The current version is MPI2. This appendix serves as a brief introduction to parallel programming using Python and MPI. Interested readers are encouraged to check the MPI4Py documentation [66] and books on MPI [29] and [73] for more details.

MPI is useful on distributed memory systems and also on a shared memory system. The distributed memory system consists of a group of nodes (containing one or more processors) connected using a high speed network. Each node is an independent entity and can communicate with other nodes using MPI. The memory cannot be shared across nodes i.e., the memory location in one node is not accessible by a process running in another node. The shared memory system consists of a group of nodes that can access the same memory location from all nodes. Shared memory systems are easier to program using OpenMP, thread programming, MPI etc. as they can be imagined as one large desktop. Distributed memory systems need MPI for node-to-node communication and can also be programmed using OpenMP or thread based programming for within node computation.

There is a large amount of literature available in print as well as online that teaches MPI and OpenMP programming [70]. Since this is about Python programming, we limit the scope of this section to programming MPI using Python. We will discuss one of the MPI wrappers for Python called MPI4Py. Before we begin the discussion on MPI4Py, we will explain the need for MPI in image processing computation.

B.2 Need for MPI in Python Image Processing

Image acquisition results in the collection of billions of voxels of 3D data. Analyzing these data serially by reading an image, processing it and then reading the next one will result in long computational time. It will cause a bottleneck, especially considering that most imaging systems are closer to real time imaging. Hence it is critical to process the images in parallel. Consider an image processing operation that takes 10 minutes to process on one CPU core. If there are 100 images to be processed, the total computation time would be 1000 minutes. Instead, if the 100 images are fed to 100 different CPU cores, the images can be processed in 10 minutes, as all images are being processed at the same time. This results in a speedup of 100X. The image processing can be completed in minutes or hours instead of days or weeks. Also, many of the image processing operations such as filtering or segmentation can easily be parallelized. Hence, when one node is computing on one image, the second node can compute on a different image without the need for communication between the two nodes. Most educational and commercial institutions have either built or purchased supercomputers or clusters. Python along with MPI4Py can be used to run image processing computation faster on these systems.

B.3 Introduction to MPI4Py

MPI4Py is a Python binding built on top of MPI versions 1 and 2. It supports point-to-point communication and collective communication of Python objects. We will discuss these communications in detail. The Python objects that can be communicated need to be picklable i.e., the Python objects can be saved using Python's pickle or cPickle modules or a numpy array.

The two modes of programming MPI are single instruction multiple data (SIMD) and single program multiple data (SPMD). In SIMD programming, the same instruction runs on each node but with different data. An image processing example of SIMD processing would be performing a filtering operation by dividing the image into sub-images and writing the result to one image file for each process. In such a case, the same instruction, filtering, is performed on each of the sub-images on different nodes. In SPMD programming, a single program containing multiple instruction runs on different nodes with different data. An example would be a different filtering operation in which the image is divided into subdivisions and filtered, but instead of writing the results to a file, one of the nodes collects the filtered image and arranges them before they are saved. In this case, most of the nodes perform the same operation of filtering, while one of the nodes performs an extra operation of collecting the output from the other nodes. Generally, SPMD operations are more common than SIMD operations. We will discuss SPMD-based programming here.

An MPI program is constructed such that the same program runs on each node. To change the behavior of the program for a specific node, a test can be made for the rank of that node (also called the node number) and provide alternate or additional instructions for that node alone.

B.4 Communicator

The communicator binds groups of processes in one MPI session. In its simplest form, an MPI program needs to have at least one communicator. In the following example, we use a communicator to obtain the size and rank of a given MPI program. The first step is to import MPI from MPI4Py. Then the size and rank can then be obtained using the Get_size() and Get_rank() Python functions.

```
from mpi4py import MPI
import sys

size = MPI.COMM_WORLD.Get_size()
rank = MPI.COMM_WORLD.Get_rank()

print("Process %d among %d"% (rank, size))
```

This Python program may be run at the command-line. Typically, in a supercomputer setting, it is submitted as a job such as a portable batch system (PBS) job. An example of such a program is shown below. In the second line of the program, the number of nodes is specified using nodes, the number of processors per node using ppn, the amount of memory per processor using pmem and the time for which the program needs to be executed using walltime. The walltime in this example is 10 minutes. In the third line of the program, the directory where the Python program and other files are located is specified. The program can be saved as a text file under the name "run.pbs". The name is arbitrary and can be replaced with any other valid file name for a text file.

```
#!/bin/bash
#PBS -l nodes=1:ppn=8,pmem=1750mb,walltime=00:10:00
```

```
cd $PBS_O_WORKDIR
module load python-epd
module load gcc ompi/gnu
mpirun -np 8 python firstmpi.py
```

The PBS script must be submitted using the command "qsub run.pbs". The queuing system completes its tasks and outputs two files: an error file containing any error messages generated during the program execution and an output file containing the content of command line output from the program. In the next few examples, the same PBS script will be used for execution with a change in the name of the Python file.

B.5 Communication

One of the important tasks of MPI is to allow communication between two different ranks or nodes as evidenced by its name "Message Passing Interface". There are many modes of communication. The most common are point-to-point and collective communication. Communication in the case of MPI generally involves transfer of data between different ranks. MPI4Py allows transfer of any pickleable Python objects or numpy arrays.

B.5.1 Point-to-Point Communication

Point-to-point communication involves passing messages or data between only two different MPI ranks or nodes. One of these ranks sends the data while the other receives it.

There are different types of send and receive functions in MPI4Py. They are:

- Blocking communication

- Nonblocking communication

- Persistent communication

In blocking communication, MPI4Py blocks the rank until the data transfer between the ranks is completed and the rank can be safely returned to the main program. Thus, no computation can be performed on the rank until the communication is complete. This mode is inefficient as the ranks are idle during data transfer. The commonly used functions for blocking communication in MPI4Py are send(), recv(), Send(), Recv() etc.

In nonblocking communication, the node transferring does not wait for the data transfer to be completed before it begins processing the next instruction. In nonblocking communication, a test is executed at the end of data transfer to ensure its success while in blocking communication, the test is the completion of data transfer. The commonly used functions for nonblocking communication in MPI4Py are isend(), irecv(), Isend(), Irecv() etc.

In some cases, the communication needs to be kept open between pairs of ranks. In such cases, persistent communication is used. It is a subset of nonblocking communication that can be kept open. It reduces the overhead in creating and closing communication if a nonblocking communication is used instead. The commonly used functions for point-to-point communication in MPI4Py are Send_init() and Recv_init().

The following program is an example of blocking communication. The rank 0 creates a pickelable Python dictionary called data that contains two key value pairs. It then sends the "data" to the second rank using send function. The destination for this data is indicated in the dest parameter. Rank 1 (under elif statement), receives the "data" using recv function. The source parameter indicates that the data needs to be received from the rank 0.

```
from mpi4py import MPI
comm = MPI.COMM_WORLD
```

```
rank = comm.Get_rank()
if rank == 0:
    data = {'a': 7, 'b': 3.14}
    comm.send(data, dest=1, tag=11)
    print "Message sent, data is: ", data
elif rank == 1:
    data = comm.recv(source=0, tag=11)
    print "Message Received, data is: ", data
```

B.5.2 Collective Communication

Collective communication allows transmission of data between multiple ranks simultaneously. This communication is a blocking communication. A few scenarios in which can be used are:

- "Broadcast" a data to all ranks

- "Scatter" a chunk of data to different ranks

- "Gather" data from all ranks

- "Reduce" data from all ranks and perform mathematical operations

In broadcast communication, the same data is copied to all the ranks. It is used to distribute an array or object that will be used by all the ranks. For example, a Python tuple can be distributed to the various ranks as data that can be used for computation.

In the scatter method, the data is broken into multiple chunks and each of these chunks is transferred to different ranks. This method can be used for breaking (say) an image into multiple parts and transferring the parts to different ranks. The ranks can then perform the same operation on the different sub-images.

In the gather method, the data from different ranks are aggregated and moved to one of the ranks. A variation of the gather method is the "allgather" method. This method collects the data from different ranks and places them in all the ranks.

In the reduce method, the data from different ranks are aggregated and placed in one of the ranks after performing reduction operations such as summation, multiplication etc. A variation of the reduce method is the "allreduce" method. This method collects the data from different ranks, performs reduction operations and places the result in all the ranks.

The program below uses broadcast communication to pass a 3-by-3 numpy array to all ranks. The numpy array containing all ones except for the central element is created in rank 0 and is broadcast using the bcast function.

```
from mpi4py import MPI
import numpy

comm = MPI.COMM_WORLD
rank = comm.Get_rank()

if rank == 0:
    data = numpy.ones((3,3))
    data[1,1] = 3.0
else:
    pass
data = comm.bcast(data, root=0)
print "rank = ",rank
print "data = ",data
```

B.6 Calculating the Value of PI

The following program combines the elements of MPI that we have illustrated so far. The various MPI programming principles that will be used in this example are MPI barrier, MPI collective communication, specifically MPI reduce, in addition to MPI ranks.

The program calculates the value of PI using the Gregory-Leibiniz series. A serial version of the program was discussed in Chapter 2, Computing using Python modules. The program execution begins with the line "if __name__". The rank and size of the program are first obtained. The total number of terms is divided across the various ranks, so that each rank receives the same number of terms. Thus, if the program has 10 ranks and the total number of terms is 1 million, each rank will compute 100,000 terms. Once the number of terms is calculated, the "calc_partial_pi" function is called. This function calculates the "partial pi" value for each rank and stores it in the variable "partialval". The MPI barrier function is called to ensure that all the ranks have completed their computation before the next line namely comm.reduce() function is executed to sum the values from various ranks and store it in the variable "finalval". Finally, the first rank prints the value of pi, namely the content of finalval.

```
from mpi4py import MPI
import sys
import numpy as np
import time

def calc_partial_pi(rank,noofterms):
    start = rank*noofterms*2+1
    lastterm = start+(noofterms-1)*2
    denominator  = np.linspace(start,lastterm,noofterms)
    numerator = np.ones(noofterms)
```

```
    for i in range(0,noofterms):
        numerator[i] =  pow(-1,i+noofterms*rank)


    # Find the ratio and sum all the fractions
    # to obtain pi value
    partialval =  sum(numerator/denominator)*4.0
    return partialval

if __name__ == '__main__':
    comm = MPI.COMM_WORLD
    rank = comm.Get_rank()
    size = MPI.COMM_WORLD.Get_size()
    totalnoterms = 1000000
    noofterms = totalnoterms/size

    partialval = calc_partial_pi(rank,noofterms)
    comm.Barrier()
    finalval = comm.reduce(partialval,op=MPI.SUM, root=0)
    if rank==0:
        print "The final value of pi is ",finalval
```

Appendix C

Introduction to ImageJ

C.1 Introduction

In all our discussions, we have used Python for image processing. There are many circumstances where it will be helpful to view the image so that it will be easy to prototype the algorithm that needs to be written in Python for processing. There are many such software programs, the most popular and powerful being ImageJ. This appendix serves as an introduction to ImageJ. Interested readers are encouraged to check the ImageJ documentation for more details at their website [43].

ImageJ is a Java based image processing software. Its popularity is due to the fact that it has an open architecture that can be extended by using Java and macros. Due to its open nature, there are many plug-ins written by scientists and experts that are available for free.

ImageJ can read and write most image formats and also specialized formats like DICOM etc. similar to Python. Due to its ability to read and write images from many formats, ImageJ is popular in various fields of science. It is used for processing radiological images, microscope images, multi-modality images etc.

ImageJ is available on most common operating systems such as Microsoft Windows, MacOSX and Linux.

C.2 ImageJ Primer

ImageJ can be installed by following the instructions in http://rsb.info.nih.gov/ij/download.html. Depending on the operating system, the methods for running ImageJ can vary. The instructions are available in the site listed above. Since ImageJ is written using Java, the interface looks the same across all operating systems, making it easier to transition from one operating system to another. Figure C.1 shows ImageJ on MacOSX.

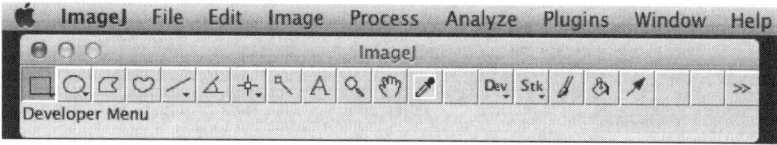

FIGURE C.1: ImageJ main screen

The files can be opened by using the File→Open menu. An example of this file is shown in Figure C.2. The 3D volume data that are stored as a series of 2D slice files can be opened using File→Import→Image Sequence... menu.

In Chapter 3, Image and its Properties, we discussed the basics of window and level. The window and level can be adjusted for the image in Figure C.2. They can be accessed using Image→Adjust→Window/Level menu and they can be adjusted by using the sliders shown in Figure C.3.

We have previously discussed various image processing techniques like filtering, segmentation etc. Such operations can also be performed using ImageJ using the Process menu. For example, the method for applying a median filter on the image is shown in Figure C.4.

Statistical information such as histogram, mean, median etc. of an image can be obtained using the Analyze menu. Figure C.5 demonstrates the method for obtaining histogram using the Analyze menu.

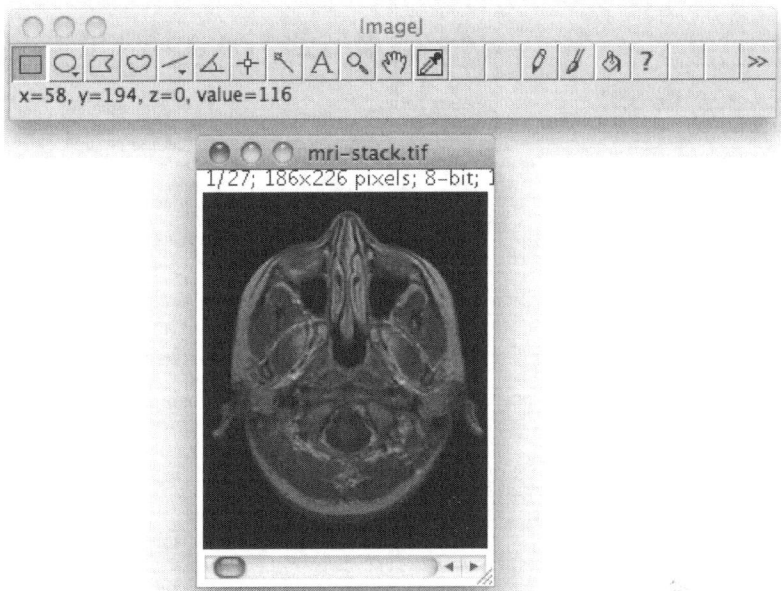

FIGURE C.2: ImageJ with an MRI image.

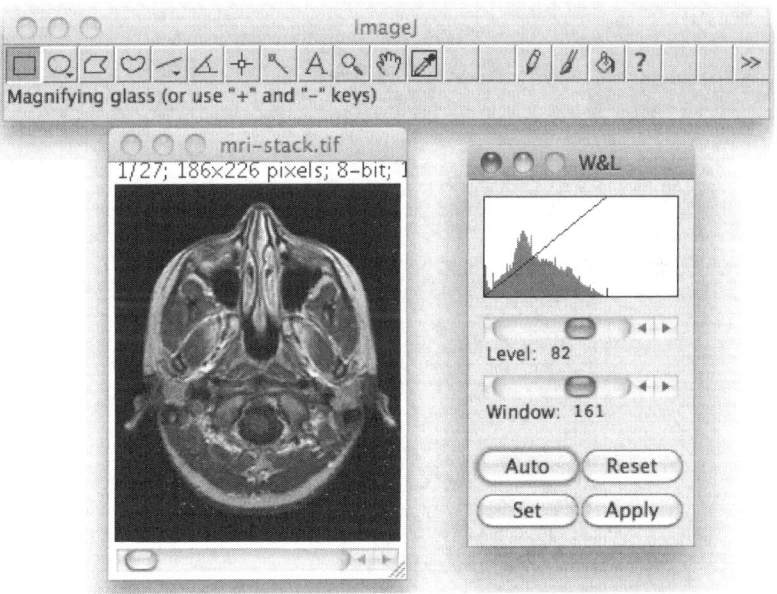

FIGURE C.3: Adjusting window or level on an MRI image.

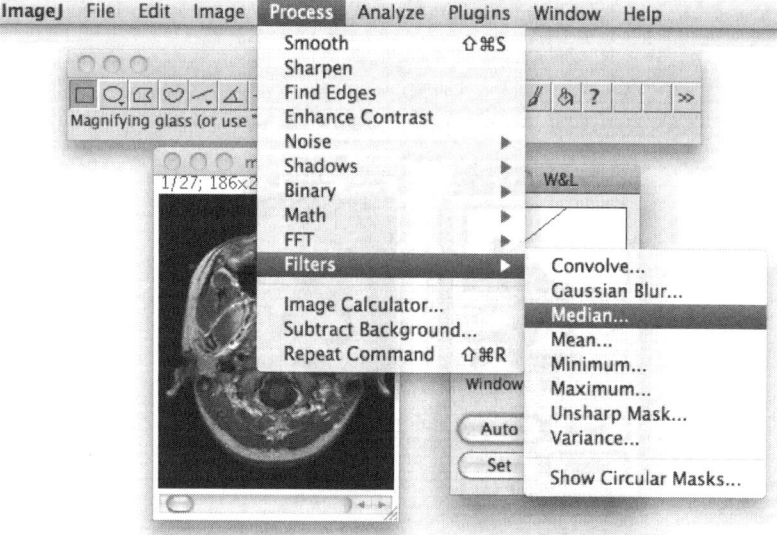

FIGURE C.4: Performing median filter.

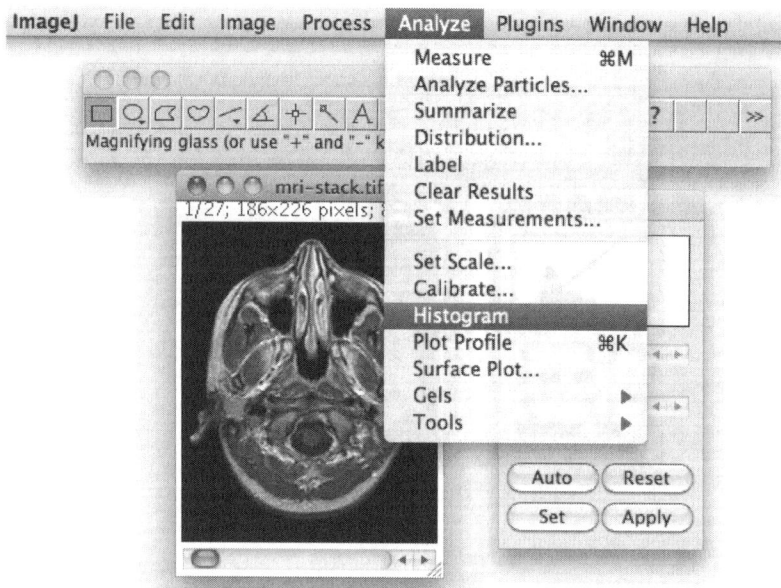

FIGURE C.5: Obtaining histogram of the image.

Appendix D

MATLAB® and Numpy Functions

D.1 Introduction

This appendix serves programmers migrating from MATLAB® to Python and interested in converting their MATLAB scripts to equivalent Python program using numpy.

MATLAB [60] is a popular commercial software that is widely used to perform computation in various fields of science including image processing. Both MATLAB and Python are interpreted languages. They both are dynamic typed, i.e., variables do not have to be declared before they are used. They both allow fast programming.

Numpy is similar in design to MATLAB in that they both operate on matrices. Because of their similarity we can find an equivalent function in MATLAB for a specific task in Numpy and vice versa. The following table lists MATLAB functions and their equivalent numpy function. The first column has numpy function, second column contains the equivalent MATLAB function, and the last column gives the description of the function. A more extensive table can be found at [89].

Numpy Function	MATLAB Equivalent	Function Description
$a[a < 10] = 0$	$a(a < 10) = 0$	Elements in a with value less than 10 are replaced with zeros.

Numpy Function	MATLAB Equivalent	Function Description
$dot(a,b)$	$a*b$	Matrix multiplication.
$a*b$	$a.*b$	Element-by-element multiplication.
$a[-1]$	$a(end)$	Access the last element in the row matrix a.
$a[1,5]$	$a(2,6)$	Access elements in columns 2 and 6 in a.
$a[3]$ or $a[3:]$	$a[4]$	Consider entire 4th row of a.
$a[0:3]$ or $a[:3]$ or $a[0:3,:]$	$a(1:3,:)$	Access first three rows of a. In Python the last index is not included in the limits.
$a[-6:]$	a(end-5:end,:)	Access the last six rows of a.
$a[0:5][:,6:11]$	$a(1:5, 7:11)$	Access row 1 to 5 and columns 7 to 11 in a.
$a[::-1,:]$	$a(end:-1:1,:)$ or $flipud(a)$	Access rows in a in reverse order.

Numpy Function	MATLAB Equivalent	Function Description
$zeros((5,4))$	$zeros(5,4)$	Array of size 5 by 4 of zeros is created. The inner circular braces are used as the size of the matrix has to be passed as a tuple.
$a[r[:len(a),0]]$	$a([1:end1,:)$	A copy of the first row will be appended at end of matrix a.
$linspace(1,2,5)$	$linspace(1,2,5)$	Five equally spaced samples between and including 1 and 2 are created.
$mgrid[0:10.,0:8.]$	$[x,y] = meshgrid(0:10,0:8)$	Creates a 2D array with x-values ranging from [0,10] and y-values ranging from [0,8].
$shape(a)$ or $a.shape$	$size(a)$	Gives the size of a.
$tile(a,(m,n))$	$repmat(a,m,n)$	Creates m by n copies of a.
$a.max()$	$max(max(a))$	Output is the maximum value in the 2D array a.
$a.transpose()$ or $a.T$	a'	Transpose of a.
$a.conj().transpose()$ or $a.conj().T$	a'	Conjugate transpose of a.
$linalg.matrix\_rank(a)$	$rank(a)$	Rank of a matrix a.

Numpy Function	MATLAB Equivalent	Function Description
$linalg.inv(a)$	$inv(a)$	Inverse of square matrix a.
$linalg.solve(a,b)$ if a is a square matrix or $linalg.lstsq(a,b)$ otherwise	a/b	Solve for x in $ax = b$.
$concatenate((a,b),1)$ or $hstack((a,b))$ or $column\_stack((a,b))$	$[a\ b]$	Concatenate columns of a and b along the horizontal direction.
$vstack((a,b))$ or $row\_stack((a,b))$	$[a;b]$	Concatenate columns of a and b along the vertical direction.

Bibliography

[1] J. Barrett and N. Keat. Artifacts in CT: Recognition and avoidance. *Radiographics*, 24(6):1679–1691, 2004.

[2] D.M. Beazley. *Python: Essential Reference*. Addison-Wesley Professional, Boston, MA, 2009.

[3] I. Bicking. Virtualenv. `http://www.virtualenv.org/en/latest/`, 2013. Accessed on 21 July 2013.

[4] W. Birkfellner. *Applied Medical Image Processing: A Basic Course*. Taylor & Francis, Boca Raton, FL, 2011.

[5] F.J. Blanco-Silva. *Learning SciPy for Numerical and Scientific Computing*. Packt Publishing, Birmingham, England, 2013.

[6] J.J. Bozzola and L.D. Russell. *Electron Microscopy, 2nd ed.* Jones & Bartlett, Burlington, MA, 1998.

[7] R.N. Bracewell. *Fourier Transform and its Applications*. McGraw-Hill, New York, NY, 1978.

[8] R.N. Bracewell. *The Impulse Symbol*. McGraw-Hill, New York, NY, 1999.

[9] E. Bressert. *SciPy and NumPy*. O'Reilly Media, Sebastopol, CA, 2012.

[10] S. Bushong. *Computed Tomography*. Essentials of medical imaging series. McGraw-Hill Education, 2000.

[11] S.C. Bushong. *Magnetic Resonance Imaging*. CV Mosby, St. Louis, MO, 1988.

[12] J. Canny. A computational approach to edge detection. *IEEE Transactions on Pattern Analysis and Machine Intelligence*, 8(6):679–698, 1986.

[13] Y. Cho, D.J. Moseley, J.H. Siewerdsen, and D.A. Jaffray. Accurate technique for complete geometric calibration of cone-beam computed tomography systems. *Medical Physics*, 32:968–983, 2005.

[14] J.W. Cooley and J.W. Tukey. An algorithm for the machine calculation of complex Fourier series. *Mathematics of Computation*, 19:297–301, 1965.

[15] T.S. Curry, J.E. Dowdey, and R.C. Murray. *Introduction to the Physics of Diagnostic Radiology*. Lea and Febiger, Philadelphia, PA, 1984.

[16] T.S. Curry, J.E. Dowdey, and R.C. Murry. *Christensen's Introduction to Physics of Diagnostic Radiology*. Lippincott Williams and Wilkins, Philadelphia, PA, 1984.

[17] cython.org. Cython module. `http://docs.cython.org/src/quickstart/build.html`, 2013. Accessed on 22 Oct 2013.

[18] C.A. Dimarzio. *Optics for engineers*. CRC Press, Boca Raton, FL, 2012.

[19] E.R. Dougherty. *Introduction to Morphological Image Processing*. SPIE International Society for Optical Engineering, 1992.

[20] D. Dowsett, P.A. Kenny, and R.E. Johnston. *The Physics of Diagnostic Imaging, 2nd ed.* CRC Press, Boca Raton, FL, 2006.

[21] M.J. Dykstra and L.E. Reuss. *Biological Electron Microscopy: Theory, Techniques, and Troubleshooting*. Kluwer Academic/Plenum Publishers, Dordrecht, The Netherlands, 2003.

[22] J. C. Elliott and S. D. Dover. X-ray microtomography. *Journal of Microscopy*, 126(2):211–213, 1982.

[23] L.C. Evans. *Partial Differential Equations, 2nd ed.* American Mathematical Society, 2010.

[24] R. Fahrig and D.W. Holdsworth. Three-dimensional computed tomographic reconstruction using a c-arm mounted xrii: Image-based correction of gantry motion nonidealities. *Medical Physics*, 27(1):30–38, 2000.

[25] L. Feldkamp, L. Davis, and J. Kress. Practical cone beam algorithm. *Journal of the Optical Society of America*, A6:612–619, 1984.

[26] D. Gilbarg and N.S. Trudinger. *Elliptic Partial Differential Equations.* Springer, New York, NY, 2001.

[27] J. Goldstein. *Scanning Electron Microscopy and X-ray Microanalysis*, volume v. 1. Kluwer Academic/Plenum Publishers, Dordrecht, The Netherlands, 2003.

[28] R.C. Gonzalez, R.E. Woods, and S.L. Eddins. *Digital image processing using MATLAB®, 2nd ed.* Gatesmark Publishing, TN, 2009.

[29] W. Gropp, E.L. Lusk, and A. Skjellum. *Using MPI, 2nd ed.* The MIT Press, Boston, MA, 1999.

[30] A.N. Hajibagheri. *Electron Microscopy: Methods and Protocols.* Humana Press, New York, NY, 1999.

[31] A. Hayat. *Principles and Techniques of Electron Microscopy: Biological Applications.* Cambridge University Press, Cambridge, England, 2000.

[32] S.L. Fleglerand J.W. Heckman and K.L. Klomparens. *Scanning*

and *Transmission Electron Microscopy: An Introduction*. Oxford University Press, Oxford, England, 1993.

[33] W.R. Hendee. *The Physical Principles of Computed Tomography*. Little, Brown library of radiology. Little Brown, New York, NY, 1983.

[34] C.L.L. Hendriks, G. Borgefors, and R. Strand. *Mathematical Morphology and Its Applications to Signal and Image Processing*. Springer, New York, NY, 2013.

[35] B. Herman and J.J. Lemasters. *Optical microscopy: Emerging Methods and Applications*. Academic Press, Waltham, MA, 1993.

[36] M.L. Hetland. *Python Algorithms: Mastering Basic Algorithms in the Python Language*. Apress, New York, NY, 2010.

[37] L. Hong, Y. Wan, and A. Jain. Fingerprint image enhancement: algorithm and performance evaluation. *IEEE Transactions on Pattern Analysis and Machine Intelligence*, 20(8):777–789, 1998.

[38] A.L. Horowitz. *MRI Physics for Radiologists: A Visual Approach*. Springer-Verlag, New York, NY, 1995.

[39] J. Hsieh. *Computed Tomography: Principles, Design, Artifacts, and Recent Advances*. SPIE, 2003.

[40] I. Idris. *NumPy Cookbook*. Packt Publishing, Birmingham, England, 2012.

[41] J. Illingworth and J. Kittler. The adaptive Hough transform. *IEEE Transactions on Pattern Analysis and Machine Intelligence*, 9(5):690–698, 1987.

[42] J. Illingworth and J. Kittler. A survey of the Hough transform. *Computer Vision, Graphics, and Image Processing*, 44(1):87–116, 1988.

[43] National Health Institue. ImageJ documentation. http://imagej.nih.gov/ij/docs/guide/, 2013. Accessed on 21 July 2013.

[44] P.M. Joseph and R.D. Spital. A method for correcting bone induced artifacts in computed tomography scanners. *Journal of Computer Assisted Tomography*, 2:100–108, 1978.

[45] A.C. Kak and M. Slaney. *Principles of Computerized Tomographic Imaging*. IEEE Press, New York, NY, 1988.

[46] W. Kalender. *Computed Tomography: Fundamentals, System Technology, Image Quality, Applications*. Publicis MCD Verlag, 2000.

[47] R.J. Keyse. *Introduction to Scanning Transmission Electron Microscopy*. Microscopy Handbooks. Bios Scientific Publishers, Oxford, England, 1997.

[48] Python Enterprise Application Kit. Easyinstall. http://peak.telecommunity.com/DevCenter/EasyInstall, 2013. Accessed on 22 Oct 2013.

[49] E. Kohl and W. Burton. *The Electron Microscope; An Introduction to Its Fundamental Principles and Applications*. Reinhold, 1946.

[50] J. Kuo. *Electron Microscopy: Methods and Protocols*. Methods in Molecular Biology. Humana Press, New York, NY, 2007.

[51] J.P. Lewis. Fast template matching. *Vision Interface*, 95:120–123, 1995.

[52] H. Li, M.A. Lavin, and R.J. Le Master. Fast hough transform: A hierarchical approach. *Computer Vision, Graphics, and Image Processing*, 36(2-3):139–161, 1986.

[53] L.A. Love and R.A. Kruger. Scatter estimation for a digital radiographic system using convolution filtering. *Medical Physics*, 14(2):178–185, 1987.

[54] M. Lutz. *Programming Python*. O'Reilly, Sebastopol, CA, 2006.

[55] A. Macovski. *Medical Imaging Systems*. Prentice Hall, Upper Saddle River, NJ, 1983.

[56] J.A. Markisz and J.P. Whalen. *Principles of MRI: Selected Topics*. Appleton & Lange, East Norwalk, CT, 1998.

[57] D. Marr and E. Hildreth. Theory of edge detection. *Proceedings of the Royal Society of London. Series B, Biological Sciences*, 207(1167):187–217, 1980.

[58] A. Martelli. Edge detection using heuristic search methods. *Computer Graphics and Image Processing*, 1(2):169–182, 1972.

[59] Materialise. Mimics^{TM}. http://biomedical.materialise.com/mimics, 2013. Accessed on 30 Oct 2013.

[60] Mathworks. Matlab®. http://www.mathworks.com/, 2013. Accessed on 21 July 2013.

[61] D.W. McRobbie. *MRI from Picture to Proton*. Cambridge University Press, Cambridge, England, 2003.

[62] J. Mertz. *Introduction to Optical Microscopy*. Roberts and Company, Greenwood Village, CO, 2010.

[63] F. Meyer. Color image segmentation. *Proceedings of the International Conference on Image Processing and its Applications*, pages 303–306, 1992.

[64] F. Meyer. Topographic distance and watershed lines. *Signal Processing*, 38:113–125, 1994.

[65] F. Meyer and S. Beucher. Morphological segmentation. *Journal of Visual Communication and Image Representation*, 1(1):21–46, 1990.

[66] MPI4Py.org. Mpi4py. `http://mpi4py.scipy.org/docs/usrman/`, 2013. Accessed on 22 Oct 2013.

[67] L. Najman and H. Talbot. *Mathematical Morphology*. Wiley-ISTE, 2010.

[68] B. Ohnesorge, T. Flohr, and K. Klingenbeck-Regn. Efficient object scatter correction algorithm for third and fourth generation CT scanners. *European Radiology*, 9:563–569, 1999.

[69] OpenCV. cv2 documentation. `http://http://docs.opencv.org`, 2013. Accessed on 21 July 2013.

[70] OpenMP.org. OpenMP. `http://openmp.org/wp/openmp-specifications/`, 2013. Accessed on 22 Oct 2013.

[71] S. Osher and L.I. Rudin. Feature-oriented image enhancement using shock filters. *SIAM Journal Numerical Analysis*, 27(4):919–940, 1989.

[72] N. Otsu. A threshold selection method from gray level histograms. *IEEE Transactions on Systems, Man and Cybernetics*, 9(1):62–66, 1979.

[73] P. Pacheco. *An Introduction to Parallel Programming*. Morgan Kaufmann, Burlington, MA, 2011.

[74] S.K. Pal and R.A. King. Image enhancement using smoothing with fuzzy sets. *IEEE Transactions on Systems, Man, and Cybernetics*, 11(7):494–501, 1981.

[75] J.R. Parker. Gray level thresholding in badly illuminated images. *IEEE Transactions on Pattern Analysis and Machine Intelligence*, 13:813–819, 1991.

[76] M. Petrou and J. Kittler. Optimal edge detectors for ramp edges. *IEEE Transactions on Pattern Analysis and Machine Intelligence*, 13(5):483–491, 1991.

[77] J.M.S. Prewitt. Object enhancement and extraction. *Picture Processing and Psychopictorics*, pages 75–149, 1970.

[78] P. Raybaut and G. Nyo. Pythonxy. http://code.google.com/p/pythonxy/, 2013. Accessed on 22 Oct 2013.

[79] A. Renyi. On measures of entropy and information. *Proceedings of Fourth Berkeley Symposium on Mathematics Statistics and Probability*, pages 547–561, 1961.

[80] G.S. Robinson. Detection and coding of edges using directional masks. *Optical Engineering*, 16(6):166580–166580, 1977.

[81] K. Rogers, P. Dowswell, K. Lane, and L. Fearn. *The Usborne Complete Book of the Microscope: Internet Linked*. Complete Books. EDC Publishing, Tulsa, OK, 2005.

[82] W. Röntgen. On a new kind of rays. *Würzburg Physical and Medical Society*, 137:132–141, 1895.

[83] J.C. Russ. *The Image Processing Handbook, 6th ed.* CRC Press, Boca Raton, FL, 2011.

[84] P.K. Sahoo, S. Soltani, and A.K.C. Wong. A survey of thresholding techniques. *Computer Vision, Graphics, and Image Processing*, 4(8):233–260, 1988.

[85] R.J. Schalkoff. *Digital Image Processing and Computer Vision*. Wiley, New York, 1989.

[86] H.M. Schey. *Div, Grad, Curl, and All That, 4th ed.* W.W. Norton and Company, New York, NY, 2004.

[87] Scikits-image.org. Scikits-image. http://scikit-image.org/docs/dev/api/skimage.measure.html, 2013. Accessed on 22 Oct 2013.

[88] Scikits.org. Scikits. http://scikit-image.org/docs/dev/api/api.html, 2013. Accessed on 22 Oct 2013.

[89] SciPy.org. Numpy to MATLAB®. http://www.scipy.org/NumPy_for_Matlab_Users, 2013. Accessed on 21 July 2013.

[90] SciPy.org. Scipy. http://docs.scipy.org/doc/scipy/reference, 2013. Accessed on 22 Oct 2013.

[91] SciPy.org. Scipy ndimage. http://docs.scipy.org/doc/scipy/reference/ndimage.html, 2013. Accessed on 22 Oct 2013.

[92] J. Serra. *Image analysis and mathematical morphology*. Academic Press, Waltham, MA, 1982.

[93] J. Serra and P. Soille. *Mathematical Morphology and Its Applications to Image Processing*. Springer, New York, NY, 1994.

[94] L. Shafarenko, H. Petrou, and J. Kittler. Histogram-based segmentation in a perceptually uniform color space. *IEEE Transactions on Image Processing*, 7(9):1354–1358, 1998.

[95] C.E. Shannon. A mathematical theory of communication. *Bell System Technical Journal*, 27:379–423, 1948.

[96] V.A. Shapiro. On the Hough transform of multi-level pictures. *Pattern Recognition*, 29(4):589–602, 1996.

[97] J.O. Smith. *Mathematics of Discrete Fourier Transform: With Audio Applications*. W3K, 2007.

[98] P. Soille. *Morphological Image Analysis: Principles and Applications, 2nd ed.* Springer, New York, NY, 2004.

[99] Enthought Scientific Computing Solutions. Enthought canopy. www.enthought.com, 2013. Accessed on 22 Oct 2013.

[100] M. Sonka, V. Hlavac, R. Boyle, et al. *Image Processing, Analysis, and Machine Vision*. PWS, Pacific Grove, CA, 1999.

[101] R. Splinter. *Handbook of Physics in Medicine and Biology*. CRC Press, Boca Raton, FL, 2010.

[102] E. Stein and R. Shakarchi. *Fourier Analysis: An Introduction*. Princeton University Press, Princeton, NJ, 2003.

[103] S. Vaingast. *Beginning Python Visualization: Crafting Visual Transformation Scripts*. Apress, New York, NY, 2009.

[104] G. Wang, D.L. Snyder, J.A. O'Sullivan, and M.W. Vannier. Iterative deblurring for CT metal artifact reduction. *IEEE Transactions on Medical Imaging*, 15:657–664, 1996.

[105] I.M. Watt. *The Principles and Practice of Electron Microscopy*. Cambridge University Press, Cambridge, England, 1997.

[106] C. Westbrook. *MRI at a Glance*. Wiley, New York, NY, 2009.

[107] L. Xu and E. Oja. Randomized Hough transform: Basic mechanisms, algorithms, and computational complexities. *Computer Vision, Graphics, and Image Processing*, 57(2):131–154, 1993.

[108] W. Zhang, S. Mukhopadhyay, S.V. Pletnev, T.S. Baker, R.J. Kuhn, and M.G. Rossmann. Placement of the structural proteins in Sindbis virus. *Journal of Virology*, 76:11645–11658, 2002.

Index

Absorption cross-section, 286
Adaptive thresholding, 149–151
Angiography, 224–226
Attentuation, 216–218
Auger electrons, AE, 298

Back-scattered electrons, BSE, 298
Bandpass filter, 134
Bloch equation, 250–251
Braking spectrum, 213
Bremsstrahlung electrons, 213
Bremsstrahlung spectrum, 213
Bremsstrahlung x-rays, 298
Butterworth highpass filter, 130
Butterworth lowpass filter, 123

Canny filter, 76–79
Catchment basin, 153
Characteristic spectrum, 213
Characteristic x-rays, 298
Computed Tomography, 226–242
    artifacts, 237–242
        beam hardening, 241–242
        central slice theorem, 228–231
        cone beam, 233–234
        dark current, 240
        fan beam, 232–233
        geometric misalignment artifacts, 238
        Hounsfield unit (HU), 236–237
        metal artifacts, 242
        micro, 234–235
        offset and gain correction, 240–241
        parallel beam, 227
        reconstruction, 227
        scatter, 238–239
Confocal microscopes, 288–289
Contrast stretching, 103–106

Dichroic mirror, 287, 289
Diffracted electrons (DE), 298
Dilation
    binary, 171
    grayscale, 175
Distance transform, 154
Dye, 284

Edge detection, 69–85
    edges, 69
Elastically scattered electrons, 298
Electromagnetic lens, 303–304
Electron beam, 296

Electron gun, 301–303
Electron microscopes, 295–309
Enthought Python Distribution (EPD), 6–7
Epi-illumination, 284
Erosion
    grayscale, 175
Everhart-Thornley detector, 304, 305
Eyepiece, 276

Faraday's law, 248–249
Field emission gun, 301
Filter
    first derivative, 71
    linear, 57
    mean, 60
    non-linear, 57
    second derivative, 79
    spatial, 57
Filters, 287–288
    bandpass, 287
    highpass, 287
    lowpass, 287
Flat Panel Detector (FPD), 223–224
Fluorescence, 284
Fluorescence microscope, 284–288
Fluorochrome, 284
Fluoroscopy, 224
Fourier transform, 109
    2D Fourier transform, 113
    2D inverse Fourier, 114
    convolution, 118
    definition, 110
    discrete FT, 111–118
    FFT, 115
    frequency domain, 109
    frequency variable, 111
    frequency variables, 114
    inverse FT, 110
Frequency encoding, 259

Gaussian highpass filter, 132
Gaussian lowpass filter, 125
Geiger Muller counter, 219
Gradient echo imaging, 267–268
Gradient magnet, 260–261
Gyromagnetic ratio, 249, 251–252

Histogram based segmentation, 139–148
Histogram equalization, 99–103
Hit-or-miss, 179–183
Hough transform, 194–200
Huygens-Fresnel principle, 278

Ideal highpass filter, 127
Ideal lowpass filter, 120
Image enhancement
    histogram equalization, 99
    image inverse, 91
    log transformation, 97
    pixel transformation, 89
    power law transformation, 92
Image intensifier, 220–221
Image inverse, 91–92

Inelastically scattered electrons, 298
Inhomogeneity artifact, 271–272
Inversion recovery, 266–267
Ionization detection, 219

Jablonski, 284

K-space imaging, 262–263
Köhler illumination, 283
Kinetic energy, 297

Lambert-Beer law, 218
Laplacian filter, 79–82
Laplacian of Gaussian, 83–85
Larmor frequency, 249
Light microscopes, 275–292
Log transformation, 97–99

Magnification, 276
Main magnet, 259–260
Marker image, 154
Max filter, 66–68
Mean filter, 60–64
Median filter, 64–66
Metal artifact, 271
Meyers flooding algorithm, 154
Min filter, 68–69
Morphology, 165–186
    closing, 176
    dilation, 166
    erosion, 171
    hit-or-miss transformation, 179
    opening, 176
    skeletonization, 185
    thickening, 184
    thinning, 184
Motion artifact, 269
MRI artifacts, 268–272
Multi-channel imaging, 287
Multiple-field II, 221–222

Nipkow disk microscopes, 289–290
Nuclear Magnetic Resonance (NMR), 247
Numerical aperture, 277–281

Objective lens, 276, 280–281
Otsu's method, 141–143

Padding, 59
Partial volume artifact, 272
Phase encoding, 258–259
Planck's constant, 213
Point spread function (PSF), 281–282
Power law transformation, 92–96
Prewitt filter, 73–76
Proton density, 252–253
Python
    comments, 9
    CSV files, 18
    data structures, 11–19
    dictionaries, 16–17
    excel files, 18–19
    file handling, 17–18
    for-loop, 10–11
    if-else statement, 11
    indentation, 8

list comprehensions, 14–16
list functions, 13
lists, 12–13
loops, 10
operators, 9
user defined functions, 19
variables, 9
Python interpreter, 6
PythonXY, 5, 7

Rayleigh criterion, 280
Refractive index, 278, 280
Region based segmentation, 151–160
Relaxation times, 253–255
Renyi entropy, 144–148
Resolution, 278, 295
　　diffraction limit, 278
　　limiting, 278
Resolving power, 279
RF coils, 261

Scanning electron microscope (SEM), 300–301
Scattered EM, SEM, 308
Scintillation detection, 220–221
Secondary electrons, SE, 298
Secondary wave sources, 278
Segmentation, 139
　　adaptive thresholding, 149
　　histogram based, 139–148
　　Otsu's method, 141
　　region based, 151
　　watershed, 153–160

Shannon entropy, 145
Skeletonization, 185–186
Slice selection, 258
Sobel filter, 76
Spatial domain, 57
Spatial filters
　　max, 66
　　mean, 60
　　median, 64
Spectral distribution, 212
Spectrum, 212
Speed of light, 214
Spin echo imaging, 265–266

Thermionic gun, 301
Trans-illumination, 284
Transmission EM, TEM, 298–300, 307
Transmitted electrons, 298

Watershed segmentation, 153–160
Wide-field microscopes, 282–283

X-ray, 209–226
　　detection, 219–224
　　generation, 210–215
　　material properties, 216–219